METHODS IN MOLECULAR BIOLOGY

For further volumes:
http://www.springer.com/series/7651

For over 35 years, biological scientists have come to rely on the research protocols and methodologies in the critically acclaimed *Methods in Molecular Biology* series. The series was the first to introduce the step-by-step protocols approach that has become the standard in all biomedical protocol publishing. Each protocol is provided in readily-reproducible step-by-step fashion, opening with an introductory overview, a list of the materials and reagents needed to complete the experiment, and followed by a detailed procedure that is supported with a helpful notes section offering tips and tricks of the trade as well as troubleshooting advice. These hallmark features were introduced by series editor Dr. John Walker and constitute the key ingredient in each and every volume of the *Methods in Molecular Biology* series. Tested and trusted, comprehensive and reliable, all protocols from the series are indexed in PubMed.

Immune Homeostasis

Methods and Protocols

Edited by

Vijay Kumar

Department of Surgery, Morehouse School of Medicine, Atlanta, GA, USA

Humana Press

Editor
Vijay Kumar
Department of Surgery
Morehouse School of Medicine
Atlanta, GA, USA

ISSN 1064-3745 ISSN 1940-6029 (electronic)
Methods in Molecular Biology
ISBN 978-1-0716-3753-1 ISBN 978-1-0716-3754-8 (eBook)
https://doi.org/10.1007/978-1-0716-3754-8

Cover Image: Mouse lung after recovering from Gram-negative bacterial pneumonia.

This Humana imprint is published by the registered company Springer Science+Business Media, LLC, part of Springer Nature.
The registered company address is: 1 New York Plaza, New York, NY 10004, U.S.A.

Paper in this product is recyclable.

Preface

Immune homeostasis deals with the homeostasis within the immune system comprising innate and adaptive immunity, which is a tightly regulated immune network. The well-developed or maintained immune system is crucial for the host defense against invading pathogens, or the microbe or pathogen-associated molecular patterns (PAMPs), allergens, developing immune tolerance to the self-antigens, and the host microbiome (skin microbiota, gut microbiota, respiratory tract microbiota, and vaginal microbiota), and recognizing host-derived death or damage-associated molecular patterns (DAMPs). People with an underdeveloped immune system or primary and secondary immunodeficiencies, including premature infants, neonates, aging population (≥60 years old), and people with genetic disorders causing immunodeficiency or suffering from infections (severely comprised immunodeficiency or SCID, human immunodeficiency virus-1 or HIV-1 induced acquired-immunodeficiency syndrome or AIDS).

The immunosuppression predisposes the host to different infections, including opportunistic infections, as seen in patients with AIDS, and the incidence of *Mycobacterium tuberculosis* (TB) is higher in the immunocompromised hosts. Cancer is also associated with disturbed immune homeostasis, causing a weak immune response against tumor cells, and the tumor microenvironment reprograms these immune cells to support immunosuppression and growth, differentiation, proliferation, and metastasis of cancer cells. On the other hand, conditions (persistent infections, molecular mimicry, exposure to self-antigens, and DAMPs) causing over-activation of the immune system may lead to autoimmune and autoinflammatory diseases. Furthermore, the activated immune system should return to its homeostasis after the clearance of potential pathogen, allergen, or inflammogen, as the failure to do this proves harmful. Hence, immune homeostasis maintenance is critical for health and longevity. However, the immune system does not work by itself in isolation, as research has shown that the nervous and immune systems control each other, and disturbance in one impacts the other. For example, neurodegenerative diseases show a breach in the host immune homeostasis or develop in response to the dysregulated immune response. Thus, the current book aims to discuss the different aspects of immune homeostasis regulation and its dysregulation during various diseases, including infections, cancers, neurodegenerative diseases, autoinflammation, autoimmune diseases, and immunodeficiency, along with immunological, genetic, cell biology, and molecular biology techniques designed to understand immune homeostasis at different stages.

Atlanta, GA, USA *Vijay Kumar*

Contents

Contributors

MAURIZIO AIELLO • *React4life, Genoa, Italy; National Research Council, IEIIT Institute, Genoa, Italy*
RAZAN M. ALAJOLEEN • *Department of Biomedical Sciences and Pathobiology, Virginia Polytechnic Institute and State University, Blacksburg, VA, USA*
MARGARET BALDINI • *Department of Cardiovascular Sciences and Center for Metabolic Disease Research, Temple University Lewis Katz School of Medicine, Philadelphia, PA, USA*
ANINDITA BHATTACHARJEE • *Department of Zoology, Molecular Cell Biology laboratory, University of Kalyani, Kalyani, India*
ARUNIMA BISWAS • *Department of Zoology, Molecular Cell Biology laboratory, University of Kalyani, Kalyani, India*
PIJUSH K. DAS • *CSIR-Indian Institute of Chemical Biology, Kolkata, India*
CAROLINA DO PRADO SERVIAN • *Departamento de Biociências e Tecnologia, Instituto de Patologia Tropical e Saúde Pública, Universidade Federal de Goiás, Goiânia, GO, Brazil*
MARY DOVER • *Department of Integrative Biology and Physiology, University of California, Los Angeles, CA, USA*
MILAN FIALA • *Department of Integrative Biology and Physiology, University of California, Los Angeles, CA, USA*
SIMONE GONÇALVES FONSECA • *Departamento de Biociências e Tecnologia, Instituto de Patologia Tropical e Saúde Pública, Universidade Federal de Goiás, Goiânia, GO, Brazil*
JULIAN J. FREEN-VAN HEEREN • *Independent Scientist, Amsterdam, the Netherlands; Immunomonitoring Services, R&D, Sanquin Diagnostic Services, Amsterdam, the Netherlands*
SHUO GENG • *Department of Biological Sciences, Virginia Tech, Blacksburg, VA, USA*
MEI-LING JIANG • *Department of Neurology, Sichuan Provincial People's Hospital, University of Electronic Science and Technology of China, Chengdu, Sichuan, China*
DOMENIQUE KIVITS • *Immunomonitoring Services, R&D, Sanquin Diagnostic Services, Amsterdam, the Netherlands*
VIJAY KUMAR • *Department of Surgery, Laboratory of Tumor Immunology and Immunotherapy, Medical Education Building-C, Morehouse School of Medicine, Atlanta, GA, USA*
LIWU LI • *Department of Biological Sciences, Virginia Polytechnic Institute and State University, Blacksburg, VA, USA*
ZIHAO LI • *Department of Neurology, Nanjing Drum Tower Hospital, Medical School and the State Key Laboratory of Pharmaceutical Biotechnology, Nanjing University, Nanjing, Jiangsu, China*
XIAOFENG LIAO • *Department of Biological Sciences, Virginia Tech, Blacksburg, VA, USA*
RUICI LIN • *Department of Biological Sciences, Virginia Tech, Blacksburg, VA, USA*
DAVID COBETA LOPEZ • *Immunomonitoring Services, R&D, Sanquin Diagnostic Services, Amsterdam, the Netherlands*
XIN M. LUO • *Department of Biomedical Sciences and Pathobiology, Virginia Polytechnic Institute and State University, Blacksburg, VA, USA*

Melika Madani • *Department of Integrative Biology and Physiology, University of California, Los Angeles, CA, USA*
Letícia Carrijo Masson • *Departamento de Biociências e Tecnologia, Instituto de Patologia Tropical e Saúde Pública, Universidade Federal de Goiás, Goiânia, GO, Brazil*
Maria Elisabetta Federica Palamà • *React4life, Genoa, Italy*
Virginia Palomares Cabeza • *Department of Immunopathology, Sanquin Research and Landsteiner Laboratory, Amsterdam, the Netherlands*
Tao Pang • *State Key Laboratory of Natural Medicines, Jiangsu Key Laboratory of Drug Screening, Jiangsu Key Laboratory of Drug Discovery for Metabolic Diseases, China Pharmaceutical University, Nanjing, China*
Ketema Paul • *Department of Integrative Biology and Physiology, University of California, Los Angeles, CA, USA*
Avi Petroff • *Faculty of Medicine, Tel Aviv University, Tel Aviv, Israel*
Irma Rensink • *Immunomonitoring Services, R&D, Sanquin Diagnostic Services, Amsterdam, the Netherlands*
Silvia Scaglione • *React4life, Genoa, Italy; National Research Council, IEIIT Institute, Genoa, Italy*
John H. Stewart IV • *Department of Surgery, Laboratory of Tumor Immunology and Immunotherapy, Medical Education Building-C, Morehouse School of Medicine, Atlanta, GA, USA*
Anja ten Brinke • *Department of Immunopathology, Sanquin Research and Landsteiner Laboratory, Amsterdam, the Netherlands*
Annelies W. Turksma • *Immunomonitoring Services, R&D, Sanquin Diagnostic Services, Amsterdam, the Netherlands*
Jing Wang • *Department of Biological Sciences, Virginia Tech, Blacksburg, VA, USA*
Chuyu Wu • *Department of Neurology, Sichuan Provincial People's Hospital, University of Electronic Science and Technology of China, Chengdu, Sichuan, China; State Key Laboratory of Natural Medicines, Jiangsu Key Laboratory of Drug Screening, Jiangsu Key Laboratory of Drug Discovery for Metabolic Diseases, China Pharmaceutical University, Nanjing, China*
Yajun Wu • *Department of Biological Sciences, Virginia Tech, Blacksburg, VA, USA*
Jun Yu • *Department of Cardiovascular Sciences and Center for Metabolic Disease Research, Temple University Lewis Katz School of Medicine, Philadelphia, PA, USA*
Cheng Zhang • *Department of Cardiovascular Sciences and Center for Metabolic Disease Research, Temple University Lewis Katz School of Medicine, Philadelphia, PA, USA*
Cun-Jin Zhang • *Department of Neurology, Sichuan Provincial People's Hospital, University of Electronic Science and Technology of China, Chengdu, Sichuan, China; Department of Neurology, Nanjing Drum Tower Hospital, Medical School and the State Key Laboratory of Pharmaceutical Biotechnology, Nanjing University, Nanjing, Jiangsu, China; The Department of Neurology, Sichuan Provincial People's Hospital, University of Electronic Science and Technology of China, Chengdu, Sichuan, China*
Yao Zhang • *Department of Biological Sciences, Virginia Tech, Blacksburg, VA, USA*

Chapter 1

Immune Homeostasis: A Novel Example of Teamwork

Vijay Kumar and John H. Stewart IV

Abstract

All living organisms must maintain homeostasis to survive, reproduce, and pass their traits on to the next generation. If homeostasis is not maintained, it can result in various diseases and ultimately lead to death. Physiologists have coined the term "homeostasis" to describe this process. With the emergence of immunology as a separate branch of medicine, the concept of immune homeostasis has been introduced. Maintaining immune homeostasis is crucial to support overall homeostasis through different immunological and non-immunological routes. Any changes in the immune system can lead to chronic inflammatory or autoimmune diseases, immunodeficiency diseases, frequent infections, and cancers. Ongoing scientific advances are exploring new avenues in immunology and immune homeostasis maintenance. This chapter introduces the concept of immune homeostasis and its maintenance through different mechanisms.

Key words Homeostasis, Immune homeostasis, Immunohomeostasis, Immunity, Inflammation

1 Introduction

The concept of homeostasis was first introduced 146 years ago in 1878, by a French physiologist Claude Bernard. He defined it as maintaining a stable internal *environment by regulating various factors within a narrow range to ensure the continuity of life* [1, 2]. The term "homeostasis" was later coined by American physiologist Walter B. Cannon in 1929 [3]. *Homeostasis* refers to the process by which an organism maintains optimal conditions for survival. This can occur naturally or with the help of external factors such as diet, exercise, or medication. An organism failing to maintain homeostasis can progress to disease and early death.

Immune homeostasis is a specific type of homeostasis that maintains overall health by controlling organ function and maintaining immunological well-being. This involves various types of immune

The original version of the chapter has been revised. A correction to this chapter can be found at https://doi.org/10.1007/978-1-0716-3754-8_18

Vijay Kumar (ed.), *Immune Homeostasis: Methods and Protocols*, Methods in Molecular Biology, vol. 2782,
https://doi.org/10.1007/978-1-0716-3754-8_1,

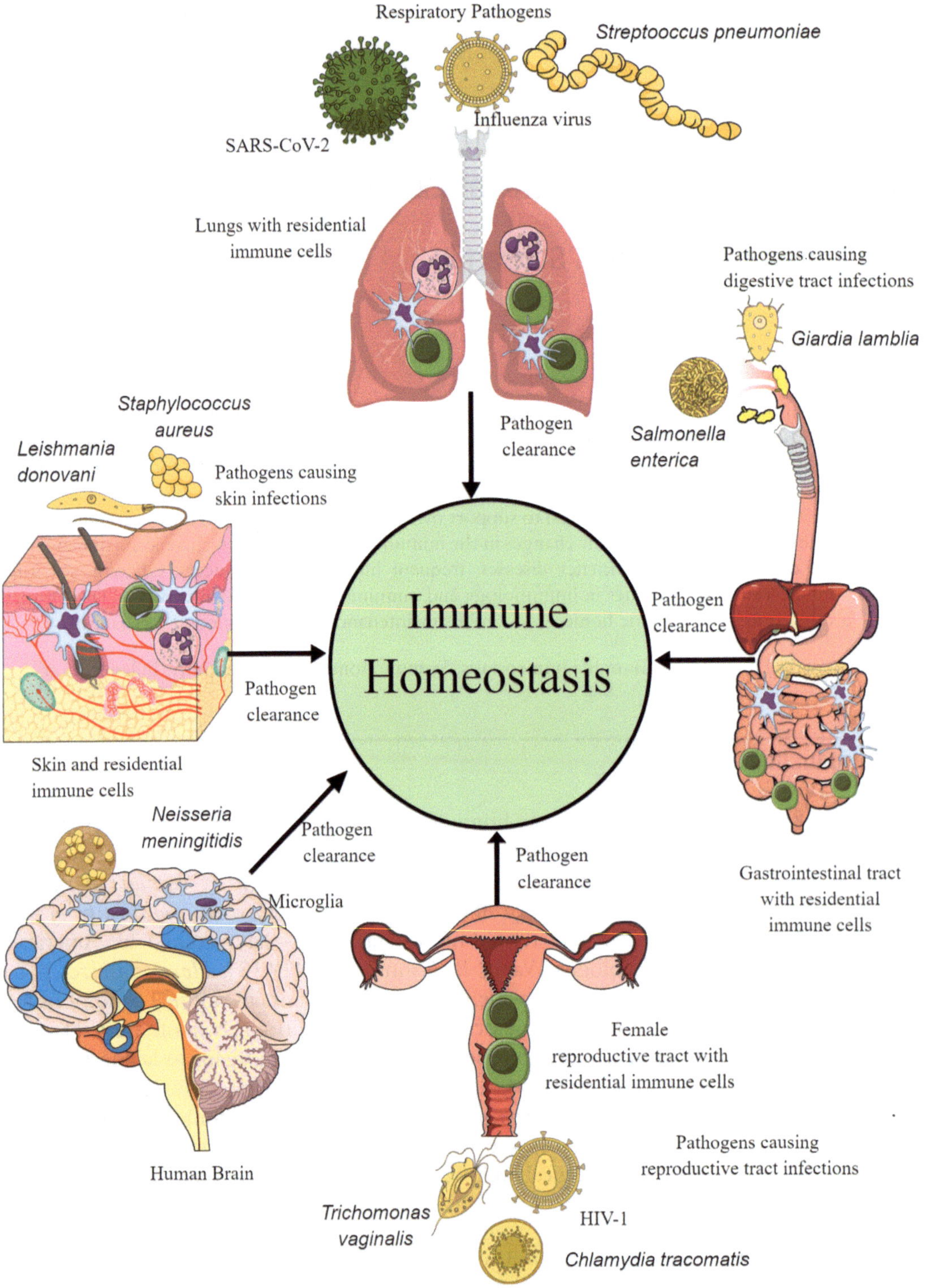

Fig. 1 Schematic representation of infection and immune homeostasis maintenance. Different microorganisms (bacteria, viruses, fungi, and parasites) invade the host through different routes, such as the respiratory tract, skin, oral or digestive tract/route, and reproductive tract. The intact organ-specific immune system recognizes the foreign pathogen and tries to clear the invading pathogen. Pathogen clearance leads to the maintenance of immune homeostasis. However, failure to clear the invading pathogen establishes the infection needing external therapeutic intervention to restore the immune homeostasis. The failure to restore

cells, such as tissue-resident macrophages, dendritic cells (DCs), innate lymphoid cells (ILCs), and tissue-resident memory T cells, which work to protect organ homeostasis (Fig. 1) [4–6]. A robust immune system can identify and combat foreign particles like microbes (such as bacteria, viruses, and fungi) and internal dangers like dying cells, modified proteins, and nucleic acids (such as DNA and RNA). This helps to maintain a healthy balance in the immune system (Fig. 1). The immune system's failure to remove foreign and internal threats predisposes the host to infectious diseases, allergies, cancers, neurodegenerative, autoimmune, immunodeficiency, and other inflammatory diseases, including metabolic diseases. Also, teaching self-tolerance prevents self-attack by the immune system for maintaining homeostasis [7, 8]. Changes to the immune system can affect the body's ability to fight disease. This can happen through various mechanisms, including immune imprinting, reprogramming, reconfiguration, and remodeling. These changes can increase the risk of developing various illnesses and put the individual in danger [9].

The coronavirus disease-2019 (COVID-19) pandemic has shown the importance of maintaining a healthy immune system, especially for those with lifestyle-related diseases and cancers and those who are older and may have immune dysregulation. This is because these groups have a higher risk of mortality from the virus [10–12]. It is essential to have a clear understanding of immune homeostasis. The immune system does not function independently but collaborates with other organ systems, such as the endocrine and nervous systems, metabolism, and microbiota (including the gut, skin, reproductive, and lung), to maintain immune homeostasis. This means that innate and adaptive immunity (also known as immune homeostasis) relies on the adaptability of organs and tissue residence to ensure a steady supply of cellular immunity throughout one's life [13]. Herein, we explore the concept of immune homeostasis and why it is crucial to understand how to maintain it. Doing so can prevent various chronic diseases, including metabolic conditions, cancer, and neurodegenerative disorders. The chapter explains how the immune system collaborates with other systems to achieve immune homeostasis.

2 Innate and Adaptive Immune Systems Collaborate to Maintain the Immune Homeostasis

The immune system consists of two parts: innate immunity and adaptive immunity. Innate immunity is made up of various types of cells, including epithelial cells, endothelial cells, neutrophils, monocytes, macrophages, DCs, mast cells, natural killer (NK) cells, basophils, eosinophils, gamma-delta ($\gamma\delta$) T cells, mucosa-associated

Fig. 1 (continued) immune homeostasis through internal defense mechanisms and external intervention leads to the ultimate loss of organ function and death of the host

invariant T (MAIT) cells, and ILCs. Additionally, the innate immune cells release humoral factors such as complement proteins, antimicrobial proteins (AMPs) like lysozymes, cathelicidins, defensins, cytokines, and chemokines to help fight infections and internal damage signals [14–16]. Adaptive immunity includes various T cells (helper and cytotoxic) and B cells that generate antibodies. In addition, there are regulatory T cells (T_{regs}), regulatory B cells (B_{regs}), and $FoxP3^{-}Helios^{+}CD8^{+}$ regulatory T cells ($CD8_{regs}$) that produce immunosuppressive and anti-inflammatory cytokines such as IL-10, IL-35, and transforming growth factor-β (TGF-β) [17–20].

The immune system is a collection of cells and their humoral factors collaborating to safeguard the body. The first line of defense is innate immunity, which combats any foreign invasion or internal threat through phagocytosis. It then generates a defensive, pro-inflammatory response to eliminate the danger. Next, innate immunity communicates with adaptive immune cells (T and B cells) through various methods, such as antigen presentation and cytokine secretion. Adaptive immune cells, in turn, secrete different factors to regulate innate immunity. Once the threat is eliminated, the pro-inflammatory immune response enters the immunosuppressive anti-inflammatory phase, known as the resolution phase. During this phase, various resolution mediators control the immune cell reprogramming, which polarizes pro-inflammatory immune cells to anti-inflammatory ones.

Pro-inflammatory cells go through apoptotic death and are eliminated by efferocytosis [21–25]. For instance, some immune cells, like pro-inflammatory M1 macrophages and N1 neutrophils, trigger a pro-inflammatory Th1 immune response. On the other hand, anti-inflammatory M2 macrophages and N2 neutrophils stimulate an anti-inflammatory Th2 immune response [26–28]. Myeloid-derived suppressor cells (MDSCs), T_{regs}, and B_{regs} help regulate the body's immune response. They prevent the development of autoimmunity or excessive inflammation by maintaining immune balance [29–35]. After the inflammation subsides, immune cells return to their normal level to maintain immune balance. However, any disruption in this carefully regulated network of immune cells, which is essential for fighting potential danger, can lead to autoimmune diseases and various cancers caused by chronic inflammation [36–39]. The tumor immune microenvironment (TIME) is filled with immune cells that have anti-inflammatory properties, such as M2 macrophages, MDSCs, N2 neutrophils, T_{regs}, and B_{regs} with specific immunometabolic reprogramming [40]. These cells help the tumor to grow [41–47] and spread by helping it to avoid immune surveillance [40–46]. The immune system maintains immune balance by combining the efforts of innate and adaptive immunity. To detect foreign substances such as microbes, pathogens, allergens, xenobiotics, and endogenous threats like death/damage-associated molecular patterns or DAMPs (proteins, DNAs, RNAs, and lipids), the immune system uses pattern recognition receptors (PRRs). These substances have specific molecular patterns known as MAMPs or PAMPs (microbe- or pathogen-associate molecular patterns).

PRRs are expressed on the extracellular cell membrane, located in the cytosol, and different cytosolic organelles such as endosomes, endolysosomes, phagosome, phagolysosomes, endoplasmic reticulum (ER), Golgi complex, and nuclear compartment (Table 1). The host uses various PRRs like toll-like receptors (TLRs), retinoic acid–inducible gene 1 (RIG-1)-like receptors (RLRs), C-type lectin receptors (CLRs), nucleotide-binding and oligomerization domain (NOD)-like receptors (*NLRs*), and absent in melanoma-2-like receptors (ALRs) to maintain a healthy immune system (Table 1). These receptors help identify and remove harmful substances like PAMPs/MAMPs and DAMPs from inside and outside the cells. Additionally, the host uses inflammasomes, cGAS/STING signaling pathway, and different complement receptors (C3aR and C5aR)/proteins to recognize and clear these substances, which ultimately helps maintain healthy immune homeostasis (Table 1) [48–56]. Various PRRs work together and communicate with each other to ensure a balanced immune response during both infections and sterile inflammatory situations [53, 55, 57, 58]. The activation or suppression of these PRRs can regulate adaptive immunity by affecting the function of innate or adaptive immune cells that express PRRs [48, 59–61].

PRRs play a critical role in many diseases, including infections, cancers, autoimmune disorders, inflammatory conditions, and neurodegenerative diseases, which can throw off the immune system's balance [52, 54, 60, 62–66]. In order to control an overactive immune response, inhibitory PRRs (iPRRs) are activated. These iPRRs suppress excessive inflammation by inducing cell death, including apoptosis and autophagy, of infected or inflamed cells. They recognize specific molecular patterns such as homeostasis-altering molecular processes/patterns or HAMPs (including commensal-associated molecular patterns or CAMPs), DAMPs, and MAMPs/PAMPs [67–74]. Macrophages play a crucial role in maintaining immune balance by clearing dead cells through efferocytosis without triggering a harmful immune response. However, an imbalance in pro-inflammatory and anti-inflammatory receptors can disrupt this balance, leading to various inflammatory conditions, such as diseases caused by infections or those that are not. To address this, targeting pro-inflammatory receptors has become a promising approach for treating inflammatory diseases, infections, and cancers [54, 75–78]. Therefore, innate immune cell activation via PRRs or other mechanisms sends signals to adaptive immunity to maintain immune homeostasis under diverse conditions.

Table 1
Different PRRs, their localization, and major ligands

PRRs	Localization	Ligands (PAMPs and DAMPs)	Origin of ligands
TLR1	Plasma membrane	Triacyl lipopeptide Soluble factors	Bacteria and mycobacteria
TLR2	Plasma membrane and endosomes	Peptidoglycan (PGN), lipoteichoic acid (LTA), lipoproteins or lipopeptides, lipoarabinomannan, glycolipids, porins, zymosan, atypical LPS, heat shock protein 70 (Hsp70), and eosinophil-derived neurotoxin (EDN)	Gram positive bacteria, mycobacteria, *S. epidermidis, Trypanosoma cruzi, Treponema maltophilum, Neisseria, Fungi, Leptospira interrogans, Porphyromonas gingivalis,* and host
TLR3	Endoplasmic reticulum (ER), endosomes, multivesicular bodies, lysosomes, and endolysosomes	dsRNA and ssRNA and synthetic analog polyinosinic-polycytidylic acid (poly I:C)	dsRNA, ssRNA, and dsDNA viruses
TLR4	Plasma membrane and endosome	LPS, Taxol, fusion protein, envelope proteins, high mobility group box 1 protein (HMG-B1), Hsp60, Hsp70, Hsp22, Hsp96 type III repeat extra domain A of fibronectin, hyaluronic acid, heparin sulfate, fibrinogen, saturated fatty acids, and fetuin-A	Gram negative bacteria, plant, respiratory syncytial virus (RSV), mouse mammary tumor virus (MMTV), *Chlamydia pneumoniae, Chlamydia trachomatis,* and host
TLR5	Plasma membrane	Flagellin	Bacteria
TLR6	Plasma membrane	Di-acyl lipopeptides and zymosan	Mycoplasma
TLR7	ER, endosomes, multivesicular bodies, lysosomes, and endolysosomes	Guanosine- and uridine-rich ssRNA, loxoribine (a guanine analog), bropirimine, ribonucleoproteins (RNPs), siRNAs, and imidazoquinoline derivatives such as resiquimod (R848) and imiquimod	Viruses (human immunodeficiency virus-1 or HIV-1, influenza virus and vesicular stomatitis virus or VSV), and synthetic compounds

(continued)

Table 1 (continued)

PRRs	Localization	Ligands (PAMPs and DAMPs)	Origin of ligands
TLR8	ER, endosomes, multivesicular bodies, lysosomes, and endolysosomes	ssRNA and RNPs	Viruses
TLR9	ER, endosomes, multivesicular bodies, lysosomes, and endolysosomes	CpG oligodeoxyneucleotide (ODN), viral double-stranded DNA (dsDNA), and hemozoin pigment	Bacteria and viruses (Herpes Simplex virus-1 and -2 or HSV-1 and -2, mouse cytomegalovirus or MCMV), malaria, and protozoa genome (*Trypanosoma cruzi*)
TLR10 (humans)	ER, endosomes, multivesicular bodies, lysosomes, and endolysosomes	dsRNA [90]	Viruses
TLR11	ER, endosomes, multivesicular bodies, lysosomes, and endolysosomes	Profilin-like protein	*Toxoplasma gondii*
TLR12	ER, endosomes, multivesicular bodies, lysosomes, and endolysosomes	Profilin-like protein	*Toxoplasma gondii*
TLR13 (mouse)	ER, endosomes, multivesicular bodies, lysosomes, and endolysosomes	23s ribosomal RNA	Bacteria
NOD-like receptor (NLR) family			
NLRA family member			
MHC class II transactivator or CIITA	Nucleus of macrophages, dendritic cells (DCs), B cells, and T cells, drives MHC II transcription and other genes critical for antigen presentation, including invariant chain and HLA-DM genes	IFN-γ is a CIITA activator	Viral infections

(continued)

Table 1
(continued)

PRRs	Localization	Ligands (PAMPs and DAMPs)	Origin of ligands
NLRB family member			
Neuronal apoptosis inhibitory protein or NAIP (in humans) and NAIP1-7 (in mice)	Cytosol of myeloid immune cells (MICs) and forms inflammasome-associated with NLRC4	Flagellin, needle component of type 3 secretory system (T3SS) of the bacteria	*Legionella pneumophila*, *Salmonella typhimurium*, *Shigella flexneri*, and *Pseudomonas* spp.
NLRC family members			
Nucleotide-binding oligomerization domain 1 (NOD1)	Cytosol and activates NF-κB	Bacterial peptides, including L-Ala-γ-D-Glu-*meso*-diaminopimelic acid (Tri-DAP)	Gram-negative and -positive bacteria
NOD2	Cytosol, activates NF-κB, autophagy, and type 1 IFN production	Muramyl dipeptide (MDP)	Gram-negative and -positive bacteria
NLR family CARD domain–containing 3 (NLRC3 or sNOD3)	Cytosol of immune cells, negative regulator of T cell activation, and lipopolysaccharide (LPS)-induced immune activation, including NF-κB activation, stimulator of interferon genes (STING), and TANK-binding kinase 1 (TBK1) signaling pathway	Recognizes viral DNA and other nucleic acids via its leucine-rich repeat (LRR) domain	Viruses
NLRC4	Cytosol, plays a role in caspase 1 activation, cell death, and phagosome maturation, and works downstream to NAIP	Rod proteins and flagellin	Bacteria
NLRC5	Shuttles between nucleus and cytosol, expressed constitutively in NK, NKT, T, and B cells, and primarily upregulates MHC I and accessory protein gene expression	IFN-γ and IFN-β are NLRC5 activators. Also, NLRP3 inflammasome activators (pathogens) overlap with NLRC5 inflammasome activation	Bacteria

(continued)

Table 1 (continued)

PRRs	Localization	Ligands (PAMPs and DAMPs)	Origin of ligands
NLR family member X 1 or NLRX1	Ubiquitous in mammals and located in the outer mitochondrial membrane and mitochondrial matrix, critical for reactive oxygen species (ROS) production, regulates antiviral immunity, autophagy, and mitophagy	Recognizes Beclin 1 and human papillomavirus 16 (HPV16) E7 protein	Viruses
NLRP family members			
NLR family pyrin-domain containing 1 or NLRP1 or CARD1 or NALP1 (first PRR discovered to form inflammasome	In humans, keratinocytes and epithelial cells, including lung epithelial cells, highly express NLRP1 in the nucleoplasm and cytosol. Motor neurons of cerebral cortex and spinal cord, seminiferous tubules, and microglia, along with other immune cells	Human NLRP1 recognizes muramyl dipeptide (MDP), viral proteases, including SARS-CoV-2 and dsRNA, whereas murine NLPRPB1 recognizes virulence factor lethal toxin (LT) of the *Bacillus anthracis*	*Toxoplasma gondii* (a parasite), viruses, and bacteria
NLRP2 or NALP2	Localized to the Golgi apparatus (GA), vesicles, and cytosol of immune cells, including astrocytes, suppresses NF-κB activity, and suppresses HLA-C expression in human trophoblasts	IFN, adenosine triphosphate (ATP), and LPS can activate NLRP2	Bacteria
NLRP3 or NALP3	Inactivated NLRP3 resides in the endoplasmic reticulum (ER) and cytosol, upon activation NLRP3 inflammasome assembles in the cytosol, mitochondria, and mitochondria-associated ER membranes. It may	Several self-DAMPs, MAMPs, PAMPs, and external molecules, such as alum, silica, aluminum hydroxide, nanoparticles, UVB rays, imiquimod, and resiquimod	Host, bacteria, viruses, and fungi

(continued)

Table 1 (continued)

PRRs	Localization	Ligands (PAMPs and DAMPs)	Origin of ligands
	also localize on or adjacent to GA to form an inflammasome. Myeloid immune cells (MICs) highly express NLRP3. T cells, B cells, endothelial cells, epithelial cells, and fibroblasts also express NLRP3		
NLRP4 or NALP4	Localized to mitochondria, and not detected in immune cells but highly expressed in the placenta, testes, spleen, pancreas, liver, lungs, kidneys, and thymus. It negatively regulates TNF-α and IL-1β-induced NF-κB activation, dsDNA and dsRNA-induced type 1 IFN signaling, and regulates autophagy during bacterial infections	ATP and nucleotides	Host and viruses
NLRP5	Localized in the GA and cytosolic vesicles, not detected in immune cells and human brain but expressed in ovaries, oocytes, testes (spermatogonia), and parathyroid gland, critical for embryogenesis	ATP and nucleotides	Host
NLRP6 or NLRC6 or PYPAF5 or PAN3	Cytosol of granulocytes, T cells, monocytes, macrophages, DCs, enterocytes or intestinal epithelial cells (IECs), and myofibroblasts. NLRP6 activation inhibits NF-κB activation and caspase 1 activation	Senses microbiota metabolites, such as taurine, LTA, and LPS. However, it directly interacts with the viral dsRNA and forms a dynamic liquid-like phase condensate or liquid-liquid phase separation (LLPS)	Bacteria and viruses

(continued)

Table 1 (continued)

PRRs	Localization	Ligands (PAMPs and DAMPs)	Origin of ligands
NLRP7 or NALP7 or PYPAF3 or PAN7 or NOD12 (only found in primates)	Localized to GA of the cells of reproductive tract (testes – spermatogonia, spermatocytes, and late spermatids) and ovaries – oocytes), extravillous trophoblasts, plasma cells, plasmacytoid DCs, macrophages, and liver NK cells. Inhibits caspase-1-depedent IL-1β secretion	ATP, LPS, Pam3CSK4, PGN, LTA, and microbial acylated lipopeptides	Gram-positive bacteria, *Mycoplasma* spp. and *Mycobacterium bovis*
NLRP9 or NALP9 or NOD6, and PAN12 (Human) and NLRP9a, b, and c isoforms are present in mice	Expressed at the inner cell membrane of intestinal cells and reproductive organs such as ovaries, oocytes, and testes. Not found in immune cells.	Short dsRNA	Viruses
NLRP10 or NALP10 or PAN5 or NOD8 or PYNOD	Localized to nucleoplasm and cytosol of cells of a variety of tissue, with a high expression in the heart, skeletal muscles, brain, and skin (keratinocytes), but not in immune cells. It is also expressed in distal colonic epithelial cells (CECs). It inhibits caspase-1-dependent IL-1β secretion. It also exhibits ATPase and GTPase activity. It lacks ligand binding LRR domain. Critical for regulating host adaptive immunity during *Candida albicans* infection and other inflammatory	ATP, damaged mitochondria (in response phospholipase C activator-3m3-FBS) independent of mitochondrial DNA (mtDNA)	Host

(continued)

Table 1 (continued)

PRRs	Localization	Ligands (PAMPs and DAMPs)	Origin of ligands
	conditions such as gut autoinflammation and skin inflammatory diseases		
NLRP12 or NALP12 or PAN6 or PYPAF7 or Monarch1	Localized to the cytosol of most MICs and highly expressed in gastrointestinal cells. It inhibits NF-κB and ERK activation pathways. Also, acts as a negative regulator of NOD2 and RIG-1 (RLR) signaling pathway	ATP, heme + PAMPs or TNF, and *Yersinia pestis*	Host and pathogens
NLRP14 or NALP14 or NOD5 or PAN8	Expressed in cytosol of testes, ovarian follicle cells, and oocytes, but not in immune cells. Important for spermatogenesis and fertilization by negatively regulating cGAS/STING and RLR-mediated downstream signaling pathways through binding TBK1	Recognizes TBK1	Host protein
Absent in melanoma-2-like receptor (ALR) AIM-2	Cytosol of hematopoietic cells and its activation produces caspase-1-dependent IL-1β and IL-18 generation and pyroptosis	Recognizes cytosolic dsDNA	Host and pathogens
cGAS/STING signaling pathway	Cytosol and nucleus, but only cytosolic cGAS acts as a PRR, and STING is an adaptor molecule that recognizes cGAMP or CDNs to generate TBK1-dependent NF-κB and IRF-3-mediated type 1 IFN and cytokine release	cGAS recognizes cytosolic dsDNA	Host and pathogens

(continued)

Table 1 (continued)

PRRs	Localization	Ligands (PAMPs and DAMPs)	Origin of ligands
RIG-1-like receptors (RLRs)	Cytosol of immune cells and most other cells. RLR activation synthesizes IRF3/IRF7/NF-κB-dependent cytokines and IFNs	Recognizes viral DNA and RNA	Viruses

3 Epigenetics Is Crucial in Immune Homeostasis Maintenance

Epigenetics is a concept that was first introduced by Conrad Waddington in 1942. This term describes a change in gene expression that does not result from changes in the DNA sequence. The cell's descendants can inherit changes in gene expression to maintain the same appearance and function [79, 80]. However, epigenetic inheritance can be transgenerational by adapting the germline epigenome [81]. The environment an individual experiences during their embryonic and childhood stages, such as exposure to various infectious agents, allergens, and nutritional status, can impact their immune system's development, memory, and regulation of immune-related genes (IRGs) in later stages [82–85]. For example, early childhood stress (ELS) alters the phagocytosis-related growth arrest-specific gene six (GAS6) expression in microglia and aggravates their pro-inflammatory phenotype later in life [84]. Also, influenza vaccination in humans epigenetically alters their monocytes and decreases chromatin accessibility to activator protein-1 (AP-1) targeted loci [86]. Vaccinating against influenza enhances the accessibility of chromatin at interferon response factor (IRF) loci in monocytes and monocyte-derived dendritic cells (mDCs). This activation leads to the expression of IFN-stimulating genes (ISGs). Such an increase in accessibility of chromatin to IRF loci in innate immune cells helps boost resistance to unrelated viral infections such as Dengue and Zika in the future [86]. Infections and vaccinations during early childhood can affect the body's ability to fight off infections in the long run. They trigger changes in the immune system that can help fight off infections.

Epigenetic mechanisms, including histone modifications, DNA methylation, and noncoding RNAs such as miRNAs, regulate these changes. They control gene expression in different immune cells, which helps maintain immune balance (Fig. 2) [80, 87–

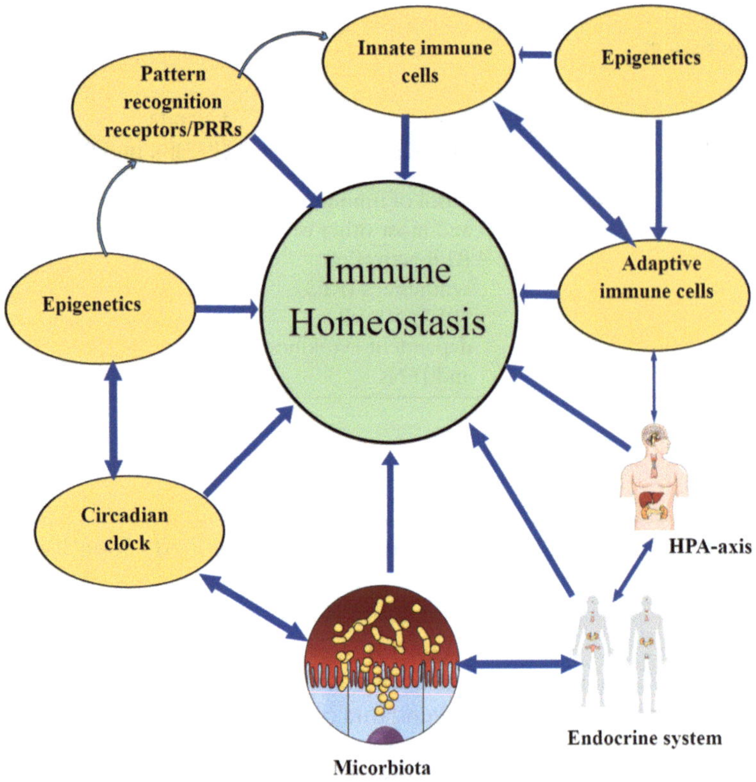

Fig. 2 Immune homeostasis needs teamwork. The immune system (innate and adaptive immune cells) works in collaboration with other systems, including neuro and endocrine systems, to maintain homeostasis. The circadian clock, nutrition, microbiota, and epigenetics work together to further impact immune homeostasis and its regulation by the HPA-axis and neuroendocrine systems. Thus, all are dependent on each other for regulating immune homeostasis

93]. Epigenetics also affects PRR expression and function, affecting immune response under diverse inflammatory and infectious diseases (Fig. 2) [94–96]. When immune regulation at the epigenetic level is disrupted, it can lead to immune tolerance issues and increase the risk of developing autoimmunity or primary immunodeficiency disorders (PIDs), such as common variable immunodeficiency (CVID). This can make the host more susceptible to frequent infections [97–101]. For example, epigenome and immune cell crosstalk are associated with naïve-to-effector-to-memory B-cell transition in the germinal center (GC)-associated epigenetic dysregulation in B cell responses in CVID and autoimmune systemic lupus erythematosus (SLE) in humans [102–105]. Additionally, epigenetic modification of immune cell (myeloid cells, T cells, and B cells) genetic network (genes) during infections, including COVID-19 and sepsis and cancers, is associated with their severity [106–109]. In addition, individuals with posttraumatic stress disorder (PTSD) or chronic stress that affects the hypothalamus-pituitary (HPA) axis may experience changes in their immune function due to alterations in their epigenome. This can result in reduced methylation of IRGs,which may suppress the immune response and increase the risk of developing immune-mediated chronic inflammatory diseases or frequent infections [110–113].

4 Immune Homeostasis Requires Teamwork

Recent developments in immunology have uncovered that maintaining immune balance relies heavily on several factors, including the circadian clock, microbiota found in the gastrointestinal, reproductive tract, lung, and skin, dietary nutrition, and endocrine and nervous systems. (Fig. 2) [114–122]. This further supports that the microbiota, endocrine system, circadian clock, nutrition, and epigenetics interaction are critical to maintaining homeostasis, and thus immune homeostasis [123–126]. For example, endocrine disrupters also disrupt microbiota to exert their toxic effect, including endocrine dysregulation [127, 128]. The gut microbiota is emerging as a virtual organ with endocrine function, giving rise to the microbiota-gut-brain axis that also affects gender-based immunity [129–134]. Gut microbiota orchestrate immune and neuroendocrine crosstalk [135]. The immune system has been found to comprise a team of players including the brain (once considered an immune-privileged organ), neurotransmitters, circadian rhythms, microbiota, and endocrine system [136–138]. A person's physical and mental well-being is closely connected to the health of their immune system. The neuroimmune axis is vital in maintaining the balance between tissue/organ and the immune system. It includes neurons and immune cells secreting neurotransmitters, cytokines, and chemokines. This system helps detect external and internal triggers and maintains homeostasis [139–144]. Furthermore, the neuroimmune axis collaborates with the host microbiota in the gut, skin, lungs, and reproductive tract to uphold immune and organ balance [145–147].

An increase in pro-inflammatory cytokines, specifically IL-1s (IL-1α and IL-1β) and IL-6, along with different prostaglandins (PGs) released from immune cells, including macrophages and endothelial cells, can cause fever (both sterile and infection induced). This can lead to illness by affecting the blood-brain barrier (BBB) and hypothalamus-pituitary (HPA) axis and inducing depression [148–153]. First, maintaining the ideal body temperature is crucial for achieving homeostasis. The hypothalamus and the immune system play essential roles in regulating body temperature [154, 155]. Additionally, pruritis, a response of the neuro-immune axis, is a natural defense mechanism against potential external or internal threats. However, chronic pruritis, as seen in individuals with atopic dermatitis, may indicate an underlying inflammatory condition caused by an imbalance in the immune system or organ function [156–161]. Nociceptor sensory neurons work with the immune system/cells to produce pain and inflammation when triggered by external and internal factors that cause inflammation [162–164]. Therefore, the nervous, endocrine, and gastrointestinal tract (GIT or gut) make many axes such as gut-brain, neuro-endocrine, and gut-endocrine to maintain immune homeostasis that also gets affected by the host's microbiota and nutritional status.

5 Conclusion

The development of various molecular, genetic, imaging, and immunological tools has led to significant progress in immunology and the regulation of immune homeostasis. One example is the microbiome, where researchers have found that the gut microbiota plays a critical role in regulating gut homeostasis and immunity through the gut-brain-immune and microbiota axis. It is essential to maintain immune homeostasis to stay healthy, as immune system dysregulation is a critical factor in many human diseases, such as allergies, infectious diseases, neurodegenerative diseases, and lifestyle-associated diseases like obesity, hypertension, and type 2 diabetes mellitus (T2DM), which are put under one umbrella called metabolic syndrome. In addition, researchers are developing therapies, such as immunomodulatory and immunotherapeutic treatments, to target infectious diseases and cancers. We anticipate that future technological advances will allow us to explore new dimensions of immune homeostasis. Hence, understanding immune homeostasis and its team players is critical for healthy living and targeting immunological disorders of various causes.

References

1. Bernard C (1879) Leçons sur les phénomènes de la vie commune aux animaux et aux végétaux, vol 2. Baillière, Paris
2. Damasio A, Damasio H (2016) Exploring the concept of homeostasis and considering its implications for economics. J Econ Behav Organ 126:125–129. https://doi.org/10.1016/j.jebo.2015.12.003
3. Cannon WB (1929) Organization for physiological homeostasis. Physiol Rev 9(3): 399–431. https://doi.org/10.1152/physrev.1929.9.3.399
4. Nobs SP, Kopf M (2021) Tissue-resident macrophages: guardians of organ homeostasis. Trends Immunol 42(6): 495–507. https://doi.org/10.1016/j.it.2021.04.007
5. Gray JI, Farber DL (2022) Tissue-resident immune cells in humans. Annu Rev Immunol 40:195–220. https://doi.org/10.1146/annurev-immunol-093019-112809
6. Gebhardt T, Palendira U, Tscharke DC, Bedoui S (2018) Tissue-resident memory T cells in tissue homeostasis, persistent infection, and cancer surveillance. Immunol Rev 283(1): 54–76. https://doi.org/10.1111/imr.12650
7. Parijs LV, Abbas AK (1998) Homeostasis and self-tolerance in the immune system: turning lymphocytes off. Science 280(5361):

243–248. https://doi.org/10.1126/science.280.5361.243

8. Chatenoud L (2014) Teaching the immune system "self" respect and tolerance. Science 344(6190):1343–1344. https://doi.org/10.1126/science.1256864
9. Halper-Stromberg A, Jabri B (2022) Maladaptive consequences of inflammatory events shape individual immune identity. Nat Immunol 23(12):1675–1686. https://doi.org/10.1038/s41590-022-01342-8
10. Dessie ZG, Zewotir T (2021) Mortality-related risk factors of COVID-19: a systematic review and meta-analysis of 42 studies and 423,117 patients. BMC Infect Dis 21(1):855. https://doi.org/10.1186/s12879-021-06536-3
11. Caramelo F, Ferreira N, Oliveiros B (2020) Estimation of risk factors for COVID-19 mortality – preliminary results. medRxiv, 2020.2002.2024.20027268. https://doi.org/10.1101/2020.02.24.20027268
12. Booth A, Reed AB, Ponzo S, Yassaee A, Aral M, Plans D et al (2021) Population risk factors for severe disease and mortality in COVID-19: a global systematic review and meta-analysis. PLoS One 16(3):e0247461. https://doi.org/10.1371/journal.pone.0247461
13. Wijeyesinghe S, Beura LK, Pierson MJ, Stolley JM, Adam OA, Ruscher R et al (2021) Expansible residence decentralizes immune homeostasis. Nature 592(7854):457–462. https://doi.org/10.1038/s41586-021-03351-3
14. Huan Y, Kong Q, Mou H, Yi H (2020) Antimicrobial peptides: classification, design, application and research progress in multiple fields [review]. Front Microbiol 11. https://doi.org/10.3389/fmicb.2020.582779
15. Kościuczuk EM, Lisowski P, Jarczak J, Strzałkowska N, Jóźwik A, Horbańczuk J et al (2012) Cathelicidins: family of antimicrobial peptides. A review. Mol Biol Rep 39(12):10957–10970. https://doi.org/10.1007/s11033-012-1997-x
16. Kumar V (2019) Inflammation research sails through the sea of immunology to reach immunometabolism. Int Immunopharmacol 73:128–145. https://doi.org/10.1016/j.intimp.2019.05.002
17. Mizoguchi A, Mizoguchi E, Takedatsu H, Blumberg RS, Bhan AK (2002) Chronic intestinal inflammatory condition generates IL-10-producing regulatory B cell subset characterized by CD1d upregulation. Immunity 16(2):219–230. https://doi.org/10.1016/s1074-7613(02)00274-1
18. Fillatreau S, Sweenie CH, McGeachy MJ, Gray D, Anderton SM (2002) B cells regulate autoimmunity by provision of IL-10. Nat Immunol 3(10):944–950. https://doi.org/10.1038/ni833
19. Mauri C, Gray D, Mushtaq N, Londei M (2003) Prevention of arthritis by interleukin 10-producing B cells. J Exp Med 197(4):489–501. https://doi.org/10.1084/jem.20021293
20. Dikiy S, Rudensky AY (2023) Principles of regulatory T cell function. Immunity 56(2):240–255. https://doi.org/10.1016/j.immuni.2023.01.004
21. Headland SE, Norling LV (2015) The resolution of inflammation: principles and challenges. Semin Immunol 27(3):149–160. https://doi.org/10.1016/j.smim.2015.03.014
22. Watanabe S, Alexander M, Misharin AV, Budinger GRS (2019) The role of macrophages in the resolution of inflammation. J Clin Invest 129(7):2619–2628. https://doi.org/10.1172/JCI124615
23. Caravaca AS, Gallina AL, Tarnawski L, Shavva VS, Colas RA, Dalli J et al (2022) Vagus nerve stimulation promotes resolution of inflammation by a mechanism that involves Alox15 and requires the α7nAChR subunit. Proc Natl Acad Sci 119(22):e2023285119. https://doi.org/10.1073/pnas.2023285119
24. Feehan KT, Gilroy DW (2019) Is resolution the end of inflammation? Trends Mol Med 25(3):198–214. https://doi.org/10.1016/j.molmed.2019.01.006
25. Fullerton JN, Gilroy DW (2016) Resolution of inflammation: a new therapeutic frontier. Nat Rev Drug Discov 15(8):551–567. https://doi.org/10.1038/nrd.2016.39
26. Mills CD (2012) M1 and M2 macrophages: oracles of health and disease. Crit Rev Immunol 32(6):463–488. https://doi.org/10.1615/critrevimmunol.v32.i6.10
27. Fridlender ZG, Sun J, Kim S, Kapoor V, Cheng G, Ling L et al (2009) Polarization of tumor-associated neutrophil phenotype by TGF-β: "N1" versus "N2" TAN. Cancer Cell 16(3):183–194. https://doi.org/10.1016/j.ccr.2009.06.017
28. Nicolás-Ávila JÁ, Adrover JM, Hidalgo A (2017) Neutrophils in homeostasis, immunity, and cancer. Immunity 46(1):15–28. https://doi.org/10.1016/j.immuni.2016.12.012

29. Mauri C, Bosma A (2012) Immune regulatory function of B cells. Annu Rev Immunol 30(1):221–241. https://doi.org/10.1146/annurev-immunol-020711-074934
30. Rosser EC, Mauri C (2015) Regulatory B cells: origin, phenotype, and function. Immunity 42(4):607–612. https://doi.org/10.1016/j.immuni.2015.04.005
31. Vignali DA, Collison LW, Workman CJ (2008) How regulatory T cells work. Nat Rev Immunol 8(7):523–532. https://doi.org/10.1038/nri2343
32. Beissert S, Schwarz A, Schwarz T (2006) Regulatory T cells. J Investig Dermatol 126(1):15–24. https://doi.org/10.1038/sj.jid.5700004
33. Dominguez-Villar M, Hafler DA (2018) Regulatory T cells in autoimmune disease. Nat Immunol 19(7):665–673. https://doi.org/10.1038/s41590-018-0120-4
34. Gabrilovich DI, Nagaraj S (2009) Myeloid-derived suppressor cells as regulators of the immune system. Nat Rev Immunol 9(3):162–174. https://doi.org/10.1038/nri2506
35. Veglia F, Sanseviero E, Gabrilovich DI (2021) Myeloid-derived suppressor cells in the era of increasing myeloid cell diversity. Nat Rev Immunol 21(8):485–498. https://doi.org/10.1038/s41577-020-00490-y
36. Medzhitov R (2008) Origin and physiological roles of inflammation. Nature 454(7203):428–435. https://doi.org/10.1038/nature07201
37. Grivennikov SI, Greten FR, Karin M (2010) Immunity, inflammation, and cancer. Cell 140(6):883–899. https://doi.org/10.1016/j.cell.2010.01.025
38. Furman D, Campisi J, Verdin E, Carrera-Bastos P, Targ S, Franceschi C et al (2019) Chronic inflammation in the etiology of disease across the life span. Nat Med 25(12):1822–1832. https://doi.org/10.1038/s41591-019-0675-0
39. Thieblemont N, Wright HL, Edwards SW, Witko-Sarsat V (2016) Human neutrophils in auto-immunity. Semin Immunol 28(2):159–173. https://doi.org/10.1016/j.smim.2016.03.004
40. Kumar V, Stewart JH (2023) Immunometabolic reprogramming, another cancer hallmark [review]. Front Immunol 14. https://doi.org/10.3389/fimmu.2023.1125874
41. Moses K, Brandau S (2016) Human neutrophils: their role in cancer and relation to myeloid-derived suppressor cells. Semin Immunol 28(2):187–196. https://doi.org/10.1016/j.smim.2016.03.018
42. Plitas G, Rudensky AY (2020) Regulatory T cells in cancer. Ann Rev Cancer Biol 4(1):459–477. https://doi.org/10.1146/annurev-cancerbio-030419-033428
43. Togashi Y, Shitara K, Nishikawa H (2019) Regulatory T cells in cancer immunosuppression – implications for anticancer therapy. Nat Rev Clin Oncol 16(6):356–371. https://doi.org/10.1038/s41571-019-0175-7
44. Balkwill F, Montfort A, Capasso M (2013) B regulatory cells in cancer. Trends Immunol 34(4):169–173. https://doi.org/10.1016/j.it.2012.10.007
45. Sarvaria A, Madrigal JA, Saudemont A (2017) B cell regulation in cancer and anti-tumor immunity. Cell Mol Immunol 14(8):662–674. https://doi.org/10.1038/cmi.2017.35
46. Barry ST, Gabrilovich DI, Sansom OJ, Campbell AD, Morton JP (2023) Therapeutic targeting of tumour myeloid cells. Nat Rev Cancer 23:216. https://doi.org/10.1038/s41568-022-00546-2
47. Groth C, Hu X, Weber R, Fleming V, Altevogt P, Utikal J et al (2019) Immunosuppression mediated by myeloid-derived suppressor cells (MDSCs) during tumour progression. Br J Cancer 120(1):16–25. https://doi.org/10.1038/s41416-018-0333-1
48. Rehwinkel J, Gack MU (2020) RIG-I-like receptors: their regulation and roles in RNA sensing. Nat Rev Immunol 20(9):537–551. https://doi.org/10.1038/s41577-020-0288-3
49. Geijtenbeek TBH, Gringhuis SI (2009) Signalling through C-type lectin receptors: shaping immune responses. Nat Rev Immunol 9(7):465–479. https://doi.org/10.1038/nri2569
50. Yamada T, Takaoka A (2023) Innate immune recognition against SARS-CoV-2. Inflamm Regen 43(1):7. https://doi.org/10.1186/s41232-023-00259-5
51. Brown GD, Willment JA, Whitehead L (2018) C-type lectins in immunity and homeostasis. Nat Rev Immunol 18(6):374–389. https://doi.org/10.1038/s41577-018-0004-8
52. Li D, Wu M (2021) Pattern recognition receptors in health and diseases. Signal Transduct Target Ther 6(1):291. https://doi.org/10.1038/s41392-021-00687-0
53. Kumar V (2019) The complement system, toll-like receptors and inflammasomes in host defense: three musketeers' one target. Int Rev

Immunol 38(4):131–156. https://doi.org/10.1080/08830185.2019.1609962
54. Kumar V (2018) Toll-like receptors in immunity and inflammatory diseases: past, present, and future. Int Immunopharmacol 59:391–412. https://doi.org/10.1016/j.intimp.2018.03.002
55. Kumar V (2020) The trinity of cGAS, TLR9, and ALRs guardians of the cellular galaxy against host-derived self-DNA. Front Immunol 11:624597. https://doi.org/10.3389/fimmu.2020.624597
56. Murugaiah V, Varghese PM, Beirag N, De Cordova S, Sim RB, Kishore U (2021) Complement proteins as soluble pattern recognition receptors for pathogenic viruses. Viruses 13(5). https://doi.org/10.3390/v13050824
57. Gradzka-Boberda S, Gentle Ian E, Häcker G (2022) Pattern recognition receptors of nucleic acids can cause sublethal activation of the mitochondrial apoptosis pathway during viral infection. J Virol 96(18):e01212–e01222. https://doi.org/10.1128/jvi.01212-22
58. Zevini A, Olagnier D, Hiscott J (2017) Crosstalk between cytoplasmic RIG-I and STING sensing pathways. Trends Immunol 38(3):194–205. https://doi.org/10.1016/j.it.2016.12.004
59. Kumar V (2022) Toll-like receptors in adaptive immunity. Handb Exp Pharmacol 276:95–131. https://doi.org/10.1007/164_2021_543
60. Jiang Y, Zhang H, Wang J, Chen J, Guo Z, Liu Y et al (2023) Exploiting RIG-I-like receptor pathway for cancer immunotherapy. J Hematol Oncol 16(1):8. https://doi.org/10.1186/s13045-023-01405-9
61. Kumar V (2021) Going, toll-like receptors in skin inflammation and inflammatory diseases. EXCLI J 20:52–79. https://doi.org/10.17179/excli2020-3114
62. Kumar V, Barrett JE (2022) Toll-like receptors (TLRs) in health and disease: an overview. Handb Exp Pharmacol 276:1–21. https://doi.org/10.1007/164_2021_568
63. Kumar V (2019) A STING to inflammation and autoimmunity. J Leukoc Biol 106(1):171–185. https://doi.org/10.1002/jlb.4mir1018-397rr
64. Kumar V (2019) Toll-like receptors in the pathogenesis of neuroinflammation. J Neuroimmunol 332:16–30. https://doi.org/10.1016/j.jneuroim.2019.03.012
65. Kigerl KA, de Rivero Vaccari JP, Dietrich WD, Popovich PG, Keane RW (2014) Pattern recognition receptors and central nervous system repair. Exp Neurol 258:5–16. https://doi.org/10.1016/j.expneurol.2014.01.001
66. Li L, Acioglu C, Heary RF, Elkabes S (2021) Role of astroglial toll-like receptors (TLRs) in central nervous system infections, injury and neurodegenerative diseases. Brain Behav Immun 91:740–755. https://doi.org/10.1016/j.bbi.2020.10.007
67. Rumpret M, von Richthofen HJ, Peperzak V, Meyaard L (2021) Inhibitory pattern recognition receptors. J Exp Med 219(1):e20211463. https://doi.org/10.1084/jem.20211463
68. Cario E, Brown D, McKee M, Lynch-Devaney K, Gerken G, Podolsky DK (2002) Commensal-associated molecular patterns induce selective toll-like receptor-trafficking from apical membrane to cytoplasmic compartments in polarized intestinal epithelium. Am J Pathol 160(1):165–173. https://doi.org/10.1016/s0002-9440(10)64360-x
69. Tukaj S, Kaminski M (2019) Heat shock proteins in the therapy of autoimmune diseases: too simple to be true? Cell Stress Chaperones 24(3):475–479. https://doi.org/10.1007/s12192-019-01000-3
70. Broere F, van der Zee R, van Eden W (2011) Heat shock proteins are no DAMPs, rather 'DAMPERs'. Nat Rev Immunol 11(8):565–565. https://doi.org/10.1038/nri2873-c1
71. Nozaki K, Li L, Miao EA (2022) Innate sensors trigger regulated cell death to combat intracellular infection. Annu Rev Immunol 40(1):469–498. https://doi.org/10.1146/annurev-immunol-101320-011235
72. Deretic V, Saitoh T, Akira S (2013) Autophagy in infection, inflammation and immunity. Nat Rev Immunol 13(10):722–737. https://doi.org/10.1038/nri3532
73. Brubaker SW, Bonham KS, Zanoni I, Kagan JC (2015) Innate immune pattern recognition: a cell biological perspective. Annu Rev Immunol 33(1):257–290. https://doi.org/10.1146/annurev-immunol-032414-112240
74. Vollmer J (2009) Autophagy links pattern recognition receptors to tumor cell apoptosis. Mol Ther 17(11):1839–1841. https://doi.org/10.1038/mt.2009.241
75. Shekarian T, Valsesia-Wittmann S, Brody J, Michallet MC, Depil S, Caux C et al (2017) Pattern recognition receptors: immune targets to enhance cancer immunotherapy. Ann Oncol 28(8):1756–1766. https://doi.org/10.1093/annonc/mdx179

76. Zhu G, Xu Y, Cen X, Nandakumar KS, Liu S, Cheng K (2018) Targeting pattern-recognition receptors to discover new small molecule immune modulators. Eur J Med Chem 144:82–92. https://doi.org/10.1016/j.ejmech.2017.12.026
77. Yuki K, Koutsogiannaki S (2021) Pattern recognition receptors as therapeutic targets for bacterial, viral and fungal sepsis. Int Immunopharmacol 98:107909. https://doi.org/10.1016/j.intimp.2021.107909
78. McKernan DP (2020) Chapter three. Pattern recognition receptors as potential drug targets in inflammatory disorders. In: Donev R (ed) Advances in protein chemistry and structural biology, vol 119. Academic Press, pp 65–109
79. Deans C, Maggert KA (2015) What do you mean, "epigenetic"? Genetics 199(4): 887–896. https://doi.org/10.1534/genetics.114.173492
80. Liotti A, Ferrara AL, Loffredo S, Galdiero MR, Varricchi G, Di Rella F et al (2022) Epigenetics: an opportunity to shape innate and adaptive immune responses. Immunology 167(4):451–470. https://doi.org/10.1111/imm.13571
81. Prokopuk L, Western PS, Stringer JM (2015) Transgenerational epigenetic inheritance: adaptation through the germline epigenome? Epigenomics 7(5):829–846. https://doi.org/10.2217/epi.15.36
82. Chen MA, LeRoy AS, Majd M, Chen JY, Brown RL, Christian LM et al (2021) Immune and epigenetic pathways linking childhood adversity and health across the lifespan. Front Psychol 12:788351. https://doi.org/10.3389/fpsyg.2021.788351
83. Fagundes CP, Glaser R, Kiecolt-Glaser JK (2013) Stressful early life experiences and immune dysregulation across the lifespan. Brain Behav Immun 27(1):8–12. https://doi.org/10.1016/j.bbi.2012.06.014
84. Reemst K, Kracht L, Kotah JM, Rahimian R, van Irsen AAS, Congrains Sotomayor G et al (2022) Early-life stress lastingly impacts microglial transcriptome and function under basal and immune-challenged conditions. Transl Psychiatry 12(1):507. https://doi.org/10.1038/s41398-022-02265-6
85. Slopen N, Loucks EB, Appleton AA, Kawachi I, Kubzansky LD, Non AL et al (2015) Early origins of inflammation: an examination of prenatal and childhood social adversity in a prospective cohort study. Psychoneuroendocrinology 51:403–413. https://doi.org/10.1016/j.psyneuen.2014.10.016
86. Wimmers F, Donato M, Kuo A, Ashuach T, Gupta S, Li C et al (2021) The single-cell epigenomic and transcriptional landscape of immunity to influenza vaccination. Cell 184(15):3915–3935.e3921. https://doi.org/10.1016/j.cell.2021.05.039
87. Xiao C, Rajewsky K (2009) MicroRNA control in the immune system: basic principles. Cell 136(1):26–36. https://doi.org/10.1016/j.cell.2008.12.027
88. Lindsay MA (2008) microRNAs and the immune response. Trends Immunol 29(7): 343–351. https://doi.org/10.1016/j.it.2008.04.004
89. Mehta A, Baltimore D (2016) MicroRNAs as regulatory elements in immune system logic. Nat Rev Immunol 16(5):279–294. https://doi.org/10.1038/nri.2016.40
90. Zhang Q, Cao X (2019) Epigenetic regulation of the innate immune response to infection. Nat Rev Immunol 19(7):417–432. https://doi.org/10.1038/s41577-019-0151-6
91. Busslinger M, Tarakhovsky A (2014) Epigenetic control of immunity. Cold Spring Harb Perspect Biol 6(6). https://doi.org/10.1101/cshperspect.a019307
92. Placek K, Schultze JL, Aschenbrenner AC (2019) Epigenetic reprogramming of immune cells in injury, repair, and resolution. J Clin Invest 129(8):2994–3005. https://doi.org/10.1172/JCI124619
93. Chen S, Yang J, Wei Y, Wei X (2020) Epigenetic regulation of macrophages: from homeostasis maintenance to host defense. Cell Mol Immunol 17(1):36–49. https://doi.org/10.1038/s41423-019-0315-0
94. Perkins DJ, Patel MC, Blanco JC, Vogel SN (2016) Epigenetic mechanisms governing innate inflammatory responses. J Interf Cytokine Res 36(7):454–461. https://doi.org/10.1089/jir.2016.0003
95. Feerick C, McKernan DP (2017) Epigenetic modifications influence NOD-like receptor expression and associated pro-inflammatory activity. FASEB J 31(S1): 1060.1062-1060.1062. https://doi.org/10.1096/fasebj.31.1_supplement.1060.2
96. Morandini AC, Santos CF, Yilmaz Ö (2016) Role of epigenetics in modulation of immune response at the junction of host-pathogen interaction and danger molecule signaling. Pathog Dis 74(7):ftw082. https://doi.org/10.1093/femspd/ftw082
97. Mehta S, Jeffrey KL (2016) Chapter 12. Immune system disorders and epigenetics. In: Tollefsbol TO (ed) Medical

epigenetics. Academic Press, Boston, pp 199–219

98. Campos-Sanchez E, Martínez-Cano J, del Pino Molina L, López-Granados E, Cobaleda C (2019) Epigenetic deregulation in human primary immunodeficiencies. Trends Immunol 40(1):49–65. https://doi.org/10.1016/j.it.2018.11.005
99. Rae W (2017) Indications to epigenetic dysfunction in the pathogenesis of common variable immunodeficiency. Arch Immunol Ther Exp 65(2):101–110. https://doi.org/10.1007/s00005-016-0414-x
100. Martínez-Cano J, Campos-Sánchez E, Cobaleda C (2019) Epigenetic priming in immunodeficiencies. Front Cell Dev Biol 7:125. https://doi.org/10.3389/fcell.2019.00125
101. Campos-Sanchez E, Martínez-Cano J, Del Pino Molina L, López-Granados E, Cobaleda C (2019) Epigenetic deregulation in human primary immunodeficiencies. Trends Immunol 40(1):49–65. https://doi.org/10.1016/j.it.2018.11.005
102. Rodríguez-Ubreva J, Arutyunyan A, Bonder MJ, Del Pino-Molina L, Clark SJ, de la Calle-Fabregat C et al (2022) Single-cell atlas of common variable immunodeficiency shows germinal center-associated epigenetic dysregulation in B-cell responses. Nat Commun 13(1):1779. https://doi.org/10.1038/s41467-022-29450-x
103. Scharer CD, Blalock EL, Mi T, Barwick BG, Jenks SA, Deguchi T et al (2019) Epigenetic programming underpins B cell dysfunction in human SLE. Nat Immunol 20(8):1071–1082. https://doi.org/10.1038/s41590-019-0419-9
104. Moroney JB, Vasudev A, Pertsemlidis A, Zan H, Casali P (2020) Integrative transcriptome and chromatin landscape analysis reveals distinct epigenetic regulations in human memory B cells. Nat Commun 11(1):5435. https://doi.org/10.1038/s41467-020-19242-6
105. Xiao F, Rui K, Shi X, Wu H, Cai X, Lui KO et al (2022) Epigenetic regulation of B cells and its role in autoimmune pathogenesis. Cell Mol Immunol 19(11):1215–1234. https://doi.org/10.1038/s41423-022-00933-7
106. Godoy-Tena G, Barmada A, Morante-Palacios O, de la Calle-Fabregat C, Martins-Ferreira R, Ferreté-Bonastre AG et al (2022) Epigenetic and transcriptomic reprogramming in monocytes of severe COVID-19 patients reflects alterations in myeloid differentiation and the influence of inflammatory cytokines. Genome Med 14(1):134. https://doi.org/10.1186/s13073-022-01137-4
107. Wu D, Shi Y, Zhang H, Miao C (2023) Epigenetic mechanisms of immune remodeling in sepsis: targeting histone modification. Cell Death Dis 14(2):112. https://doi.org/10.1038/s41419-023-05656-9
108. Belk JA, Daniel B, Satpathy AT (2022) Epigenetic regulation of T cell exhaustion. Nat Immunol 23(6):848–860. https://doi.org/10.1038/s41590-022-01224-z
109. Tien F-M, Lu H-H, Lin S-Y, Tsai H-C (2023) Epigenetic remodeling of the immune landscape in cancer: therapeutic hurdles and opportunities. J Biomed Sci 30(1):3. https://doi.org/10.1186/s12929-022-00893-0
110. Palma-Gudiel H, Prather AA, Lin J, Oxendine JD, Guintivano J, Xia K et al (2021) HPA axis regulation and epigenetic programming of immune-related genes in chronically stressed and non-stressed mid-life women. Brain Behav Immun 92:49–56. https://doi.org/10.1016/j.bbi.2020.11.027
111. Chatzittofis A, Boström ADE, Ciuculete DM, Öberg KG, Arver S, Schiöth HB et al (2021) HPA axis dysregulation is associated with differential methylation of CpG-sites in related genes. Sci Rep 11(1):20134. https://doi.org/10.1038/s41598-021-99714-x
112. Smith AK, Conneely KN, Kilaru V, Mercer KB, Weiss TE, Bradley B et al (2011) Differential immune system DNA methylation and cytokine regulation in post-traumatic stress disorder. Am J Med Genet B Neuropsychiatr Genet 156b(6):700–708. https://doi.org/10.1002/ajmg.b.31212
113. Uddin M, Aiello AE, Wildman DE, Koenen KC, Pawelec G, de los Santos R et al (2010) Epigenetic and immune function profiles associated with posttraumatic stress disorder. Proc Natl Acad Sci U S A 107(20):9470–9475. https://doi.org/10.1073/pnas.0910794107
114. Scheiermann C, Kunisaki Y, Frenette PS (2013) Circadian control of the immune system. Nat Rev Immunol 13(3):190–198. https://doi.org/10.1038/nri3386
115. Scheiermann C, Gibbs J, Ince L, Loudon A (2018) Clocking in to immunity. Nat Rev Immunol 18(7):423–437. https://doi.org/10.1038/s41577-018-0008-4
116. Pearson JA, Wong FS, Wen L (2020) Crosstalk between circadian rhythms and the microbiota. Immunology 161(4):278–290. https://doi.org/10.1111/imm.13278
117. Taylor L, Von Lendenfeld F, Ashton A, Sanghani H, Di Pretoro S, Usselmann L et al (2023) Sleep and circadian rhythm

disruption alters the lung transcriptome to predispose to viral infection. iScience 26(2): 105877. https://doi.org/10.1016/j.isci.2022.105877
118. Crislip GR, Johnston JG, Douma LG, Costello HM, Juffre A, Boyd K et al (2021) Circadian rhythm effects on the molecular regulation of physiological systems. Compr Physiol 12(1):2769–2798. https://doi.org/10.1002/cphy.c210011
119. Kiess W, Belohradsky BH (1986) Endocrine regulation of the immune system. Klin Wochenschr 64(1):1–7. https://doi.org/10.1007/bf01721574
120. Webster JI, Tonelli L, Sternberg EM (2002) Neuroendocrine regulation of immunity. Annu Rev Immunol 20(1):125–163. https://doi.org/10.1146/annurev.immunol.20.082401.104914
121. Arrieta M-C, Finlay B (2012) The commensal microbiota drives immune homeostasis [mini review]. Front Immunol 3. https://doi.org/10.3389/fimmu.2012.00033
122. Honda K, Littman DR (2016) The microbiota in adaptive immune homeostasis and disease. Nature 535(7610):75–84. https://doi.org/10.1038/nature18848
123. Bishehsari F, Voigt RM, Keshavarzian A (2020) Circadian rhythms and the gut microbiota: from the metabolic syndrome to cancer. Nat Rev Endocrinol 16(12):731–739. https://doi.org/10.1038/s41574-020-00427-4
124. Sato T, Sassone-Corsi P (2022) Nutrition, metabolism, and epigenetics: pathways of circadian reprogramming. EMBO Rep 23(5): e52412. https://doi.org/10.15252/embr.202152412
125. Coulson RL, LaSalle JM (2018) Epigenetics of circadian rhythms in imprinted neurodevelopmental disorders. In: Grayson DR (ed) Progress in molecular biology and translational science, vol 157. Academic Press, pp 67–92
126. Masri S, Kinouchi K, Sassone-Corsi P (2015) Circadian clocks, epigenetics, and cancer. Curr Opin Oncol 27(1):50–56. https://doi.org/10.1097/cco.0000000000000153
127. Aguilera M, Gálvez-Ontiveros Y, Rivas A (2020) Endobolome, a new concept for determining the influence of microbiota disrupting chemicals (MDC) in relation to specific endocrine pathogenesis [review]. Front Microbiol 11. https://doi.org/10.3389/fmicb.2020.578007
128. Williams CL, Garcia-Reyero N, Martyniuk CJ, Tubbs CW, Bisesi JH (2020) Regulation of endocrine systems by the microbiome: perspectives from comparative animal models. Gen Comp Endocrinol 292:113437. https://doi.org/10.1016/j.ygcen.2020.113437
129. Morris DJ, Ridlon JM (2017) Glucocorticoids and gut bacteria: "The GALF Hypothesis" in the metagenomic era. Steroids 125:1–13. https://doi.org/10.1016/j.steroids.2017.06.002
130. Cussotto S, Sandhu KV, Dinan TG, Cryan JF (2018) The neuroendocrinology of the microbiota-gut-brain axis: a behavioural perspective. Front Neuroendocrinol 51:80–101. https://doi.org/10.1016/j.yfrne.2018.04.002
131. Jaggar M, Rea K, Spichak S, Dinan TG, Cryan JF (2020) You've got male: sex and the microbiota-gut-brain axis across the lifespan. Front Neuroendocrinol 56:100815. https://doi.org/10.1016/j.yfrne.2019.100815
132. Yurkovetskiy L, Burrows M, Khan AA, Graham L, Volchkov P, Becker L et al (2013) Gender bias in autoimmunity is influenced by microbiota. Immunity 39(2): 400–412. https://doi.org/10.1016/j.immuni.2013.08.013
133. Chen KL, Madak-Erdogan Z (2016) Estrogen and microbiota crosstalk: should we pay attention? Trends Endocrinol Metab 27(11): 752–755. https://doi.org/10.1016/j.tem.2016.08.001
134. Bistoletti M, Bosi A, Banfi D, Giaroni C, Baj A (2020) Chapter two. The microbiota-gut-brain axis: focus on the fundamental communication pathways. In: Kasselman LJ (ed) Progress in molecular biology and translational science, vol 176. Academic Press, pp 43–110
135. El Aidy S, Dinan TG, Cryan JF (2015) Gut microbiota: the conductor in the orchestra of immune–neuroendocrine communication. Clin Ther 37(5):954–967. https://doi.org/10.1016/j.clinthera.2015.03.002
136. Wensveen FM, Šestan M, Turk Wensveen T, Polić B (2019) 'Beauty and the beast' in infection: how immune–endocrine interactions regulate systemic metabolism in the context of infection. Eur J Immunol 49(7):982–995. https://doi.org/10.1002/eji.201847895
137. Anderson AC, Acharya N (2022) Steroid hormone regulation of immune responses in cancer. Immunometabolism 4(4):e00012
138. Paavonen T (1994) Hormonal regulation of immune responses. Ann Med 26(4): 255–258. https://doi.org/10.3109/07853899409147900

139. Klein Wolterink RGJ, Wu GS, Chiu IM, Veiga-Fernandes H (2022) Neuroimmune interactions in peripheral organs. Annu Rev Neurosci 45:339–360. https://doi.org/10.1146/annurev-neuro-111020-105359
140. Godinho-Silva C, Cardoso F, Veiga-Fernandes H (2019) Neuro-immune cell units: a new paradigm in physiology. Annu Rev Immunol 37:19–46. https://doi.org/10.1146/annurev-immunol-042718-041812
141. Chesné J, Cardoso V, Veiga-Fernandes H (2019) Neuro-immune regulation of mucosal physiology. Mucosal Immunol 12(1):10–20. https://doi.org/10.1038/s41385-018-0063-y
142. Chu C, Artis D, Chiu IM (2020) Neuro-immune interactions in the tissues. Immunity 52(3):464–474. https://doi.org/10.1016/j.immuni.2020.02.017
143. Meixiong J, Basso L, Dong X, Gaudenzio N (2020) Nociceptor–mast cell sensory clusters as regulators of skin homeostasis. Trends Neurosci 43(3):130–132. https://doi.org/10.1016/j.tins.2020.01.001
144. Veiga-Fernandes H, Artis D (2018) Neuronal–immune system cross-talk in homeostasis. Science 359(6383):1465–1466. https://doi.org/10.1126/science.aap9598
145. Yu LW, Agirman G, Hsiao EY (2022) The gut microbiome as a regulator of the neuroimmune landscape. Annu Rev Immunol 40:143–167. https://doi.org/10.1146/annurev-immunol-101320-014237
146. Jacobson A, Yang D, Vella M, Chiu IM (2021) The intestinal neuro-immune axis: crosstalk between neurons, immune cells, and microbes. Mucosal Immunol 14(3):555–565. https://doi.org/10.1038/s41385-020-00368-1
147. Domingues RG, Hepworth MR (2020) Immunoregulatory sensory circuits in Group 3 innate lymphoid cell (ILC3) function and tissue homeostasis. Front Immunol 11:116. https://doi.org/10.3389/fimmu.2020.00116
148. Konsman JP, Parnet P, Dantzer R (2002) Cytokine-induced sickness behaviour: mechanisms and implications. Trends Neurosci 25(3):154–159. https://doi.org/10.1016/s0166-2236(00)02088-9
149. McCusker RH, Kelley KW (2013) Immune-neural connections: how the immune system's response to infectious agents influences behavior. J Exp Biol 216(Pt 1):84–98. https://doi.org/10.1242/jeb.073411
150. Dantzer R, O'Connor JC, Freund GG, Johnson RW, Kelley KW (2008) From inflammation to sickness and depression: when the immune system subjugates the brain. Nat Rev Neurosci 9(1):46–56. https://doi.org/10.1038/nrn2297
151. Dantzer R (2006) Cytokine, sickness behavior, and depression. Neurol Clin 24(3):441–460. https://doi.org/10.1016/j.ncl.2006.03.003
152. Leon LR (2002) Invited review: cytokine regulation of fever: studies using gene knockout mice. J Appl Physiol 92(6):2648–2655. https://doi.org/10.1152/japplphysiol.01005.2001
153. Blomqvist A, Engblom D (2018) Neural mechanisms of inflammation-induced fever. Neuroscientist 24(4):381–399. https://doi.org/10.1177/1073858418760481
154. Shao L, Pang N, Yan P, Jia F, Sun Q, Ma W et al (2018) Control of body temperature and immune function in patients undergoing open surgery for gastric cancer. Bosn J Basic Med Sci 18(3):289–296. https://doi.org/10.17305/bjbms.2018.2552
155. Evans SS, Repasky EA, Fisher DT (2015) Fever and the thermal regulation of immunity: the immune system feels the heat. Nat Rev Immunol 15(6):335–349. https://doi.org/10.1038/nri3843
156. Wang F, Kim BS (2020) Itch: a paradigm of neuroimmune crosstalk. Immunity 52(5):753–766. https://doi.org/10.1016/j.immuni.2020.04.008
157. Mack MR, Kim BS (2018) The itch–scratch cycle: a neuroimmune perspective. Trends Immunol 39(12):980–991. https://doi.org/10.1016/j.it.2018.10.001
158. Trier AM, Kim BS (2018) Cytokine modulation of atopic itch. Curr Opin Immunol 54:7–12. https://doi.org/10.1016/j.coi.2018.05.005
159. Solinski HJ, Kriegbaum MC, Tseng P-Y, Earnest TW, Gu X, Barik A et al (2019) Nppb neurons are sensors of mast cell-induced itch. Cell Rep 26(13):3561–3573.e3564. https://doi.org/10.1016/j.celrep.2019.02.089
160. Oetjen LK, Mack MR, Feng J, Whelan TM, Niu H, Guo CJ et al (2017) Sensory neurons co-opt classical immune signaling pathways to mediate chronic itch. Cell 171(1):217–228.e213. https://doi.org/10.1016/j.cell.2017.08.006
161. Hashimoto T, Yosipovitch G (2019) Itching as a systemic disease. J Allergy Clin Immunol 144(2):375–380. https://doi.org/10.1016/j.jaci.2019.04.005

162. Pinho-Ribeiro FA, Verri WA Jr, Chiu IM (2017) Nociceptor sensory neuron-immune interactions in pain and inflammation. Trends Immunol 38(1):5–19. https://doi.org/10.1016/j.it.2016.10.001

163. Baral P, Udit S, Chiu IM (2019) Pain and immunity: implications for host defence. Nat Rev Immunol 19(7):433–447. https://doi.org/10.1038/s41577-019-0147-2

164. Cook AD, Christensen AD, Tewari D, McMahon SB, Hamilton JA (2018) Immune cytokines and their receptors in inflammatory pain. Trends Immunol 39(3):240–255. https://doi.org/10.1016/j.it.2017.12.003

Chapter 2

Immune Homeostasis Maintenance Through Advanced Immune Therapeutics to Target Atherosclerosis

Shuo Geng, Yajun Wu, and Liwu Li

Abstract

Atherosclerosis remains the leading cause of coronary heart disease (CHD) with enormous health and societal tolls. Traditional drug development approaches have been focused on small molecule–based compounds that aim to lower plasma lipids and reduce systemic inflammation, two primary causes of atherosclerosis. However, despite the widely available lipid-lowering and anti-inflammatory small compounds and biologic agents, CHD prevalence still remains high. Based on recent advances revealing disrupted immune homeostasis during atherosclerosis pathogenesis, novel strategies aimed at rejuvenating immune homeostasis with engineered immune leukocytes are being developed. This chapter aims to assess basic and translational efforts on these emerging strategies for the effective development of atherosclerosis treatment, as well as key challenges in this important translational field.

Key words Immune therapeutics, Atherosclerosis, Immune cell programming, Immune modulators, Cellular dynamics, Inflammation resolution

Abbreviations

4-PBA	4-phenylbutyric acid
AHA	American Heart Association
ALOX15	Arachidonate 15-lipoxygenase
CANTOS	Canakinumab Anti-inflammatory Thrombosis Outcome Study
CD200R	Cluster of differentiation 200 receptor
CHD	Coronary heart disease
COX1	Cyclooxygenase 1
COX2	Cyclooxygenase 2
CRP	C-reactive protein
CVD	Cardiovascular disease
EVs	Extracellular vehicles
HMG-CoA	3-hydroxy-3-methylglutaryl coenzyme A
ICAM-1	Intercellular adhesion molecule 1
IL	Interleukin
LDL-C	Low-density lipoprotein cholesterol

Vijay Kumar (ed.), *Immune Homeostasis: Methods and Protocols*, Methods in Molecular Biology, vol. 2782,
https://doi.org/10.1007/978-1-0716-3754-8_2,

LDLR	Low-density lipoprotein receptor
LPS	Lipopolysaccharide
LTB4	Leukotriene B4
MI	Myocardial infarction
MPO	Myeloperoxidase
mTOR	Mammalian target of rapamycin
MyD88	Myeloid differentiation factor 88
NSAID	Nonsteroidal anti-inflammatory drug
PCSK9	Proprotein convertase subtilisin/kexin type 9
PEX5	Peroxisome biogenesis factor 5
PPARγ	Peroxisome proliferator-activated receptor gamma
ROS	Reactive oxygen species
RvD1	Resolvin D1
RvE1	Resolvin E1
STAT	Signal transducer and activator of transcription
TLR4	Toll-like receptor 4
TRAM	Toll/IL-1R domain–containing adaptor–inducing IFN-β-related adaptor molecule
TRIF	Toll/IL-1R domain–containing adaptor–inducing IFN-β

1 Introduction

Atherosclerosis-related diseases remain one of the most prevalent causes of mortality and morbidity in the world [1, 2]. The pathogenesis of atherosclerosis is highly complex and involves cellular and tissue dysfunctions at multiple levels and stages of disease progression including endothelial dysfunction, excessive infiltration of immune cells with sustained inflammation, and dysfunctional clearance of damaged debris, collectively leading to intimal thickening and atherosclerotic plaque formation. As the disease progresses, dynamic alterations of the local immune environment further orchestrate the dysregulation of plaque tissues, leading to thrombus formation and plaque rupture.

Decades of basic and clinical studies have focused on metabolic dysregulation during the initiation and progression of atherosclerosis. Elevated plasma lipids, such as low-density lipoprotein cholesterol (LDL-C) and free cholesterol, are major metabolic risk factors for atherosclerosis and related coronary heart disease (CHD) [3]. A large multicenter clinical trial in multiple countries reported that dyslipidemia accounted for nearly 50% of the risk for atherosclerosis developing myocardial infarction (MI) [4]. Furthermore, an independent trial documented that lowering LDL-C reduced cardiovascular disease (CVD) in a dose-dependent relationship [5]. Therefore, clinical guidelines for treating atherosclerosis have been largely focused on managing plasma levels of LDL-C and/or free cholesterol.

Besides elevated plasma cholesterol levels, recent compelling evidence reveals a strong connection between inflammation and the development of atherosclerosis. Systemic chronic inflammation contributes to the buildup of atherosclerotic plaques by disrupting endothelial as well as smooth muscle homeostasis, promoting immune cell entry, and compromising clearance of cellular and lipid debris [6, 7]. Large clinical trials, such as the CANTOS (Canakinumab Anti-inflammatory Thrombosis Outcome Study) trial, have been ongoing which examine the effects of anti-inflammatory drugs (e.g., canakinumab in blocking interleukin-1β and IL-1β) in combination with lipid-lowering regimens [7, 8].

Side effects associated with chemical or biologic-based therapies include off-target effects and limited delivery efficiency to inflamed atherosclerosis tissues. Recently, immune cell–based approaches have been explored that capitalize on the natural tropism of immune cells to inflamed areas of atherosclerosis. We will discuss some of the recent advances in this promising area, in the broader context of immune therapeutic development targeting atherosclerosis.

2 Molecule-Based Immune Therapeutics

Small-compound inhibitors have drawn mainstream attention, dominating the field of drug development for the treatment of atherosclerosis. Decades of studies have revealed multiple pathways and signaling molecules involved in propagating chronic inflammation during atherosclerosis, amenable for targeted drug development.

One of the earliest developed and most widely used anti-inflammatory drugs for cardiovascular disease prevention and treatment is the nonsteroidal anti-inflammatory drug (NSAID) aspirin. Inspired by naturally occurring salicylate compounds from willow tree, aspirin was developed and marketed by Beyer Pharmaceutics in the late 1800s. Aspirin suppresses the production of inflammatory lipid mediators, such as prostaglandins and thromboxanes, by inhibiting cyclooxygenases (COX1 and COX2) [9]. Low doses of aspirin redirect the enzymatic activities of COX2 to produce resolving lipid mediators such as resolvins. The American Heart Association (AHA) recommends the usage of low-dose aspirin (75–100 mg/day) for the prevention of CVD in adults aged 40–70 with higher CVD risks [10]. However, clinical studies reveal side effects of aspirin usage, including the risks of bleeding, and current guidelines caution against its usage for older populations.

Rapamycin is another widely used anti-inflammatory compound derived from microbial sources [11]. Rapamycin exerts its potent anti-inflammatory effects by inhibiting mTOR, promoting autophagy, and reducing inflammatory polarization of innate leukocytes [12, 13]. Animal studies suggest that rapamycin treatment can reduce vascular inflammation and pathogenesis of atherosclerosis [11, 14]. However, rapamycin usage can potentially exacerbate hyperlipidemia and insulin resistance, which are compound factors for atherosclerosis [15].

Since elevated levels of blood cholesterol closely correlate with human CVD risks, lipid-lowering drugs have been developed to treat cardiovascular disease. Among the most successful examples of small molecule–based lipid-lowering drugs for treating atherosclerosis are statin family molecules. Statins exert their lipid-lowering effects by suppressing the 3-hydroxy-3-methylglutaryl coenzyme A (HMG-CoA) reductase, which is involved in cholesterol synthesis [16]. A double-blind clinical trial published in the New England Journal of Medicine reported the coronary event–lowering benefits of the statin lovastatin in CVD patients [17]. The study concluded that lovastatin treatment reduced LDL-C levels, leading to a decrease in the risk of cardiovascular events. Independent analyses later suggested that statin usage not only reduced plasma cholesterol levels but also reduced key inflammatory indicators such as C-reactive protein (CRP) [18]. Despite their potent lipid-lowering and potential anti-inflammatory effects, statins do have noted side effects, such as low blood platelet count, digestive problems, and muscle pain [19]. Molecular and genetic studies reveal that proprotein convertase subtilisin/kexin type 9 (PCSK9) is an important molecule involved in modulating LDL-C levels [20, 21]. PCSK9 interacts with low-density lipoprotein receptor (LDLR) and targets LDLR for lysosomal degradation, leading to elevated levels of circulating LDL-C [22]. Missense mutations of PCSK9 in humans that yield a gain of function were observed to be responsible for a rare form of hypercholesterolemia [23]. Later, independent studies showed that patients bearing PCSK9 genetic variants exhibit atherogenic lipid profiles and increased risk of cardiovascular disease [24]. Emerging studies suggest that PCSK9 not only elevates plasma cholesterol levels but also contributes to oxidative damage and inflammation, collectively exacerbating atherosclerosis pathogenesis [25–27]. Biochemical analyses suggest that PCSK9 bears structural homology to plasma protein resistin, which may enable its binding with TLR4 and initiate inflammatory processes [28]. Inhibitors targeting PCSK9 have been developed and shown to have beneficial effects in managing lipid levels in CVD patients [29]. Biologic-based PCSK9 inhibition approaches have also been developed, which include monoclonal antibodies targeting PCSK9 approved by the FDA [30]. Although less common, rare side effects of PCSK9 inhibitors do occur which include fatigue, muscle pain, and/or potential liver/kidney problems [31].

3 Cell-Based Immune Therapeutics

The lack of delivery precision into inflamed tissues and limited efficacy are some of the intrinsic drawbacks associated with all chemical and/or biologic-based inhibitor approaches in atherosclerosis. Recent studies focusing on immune cell–based therapies may offer promising and highly effective alternatives. Immune cells are naturally equipped with chemokine receptors serving as physiological guidance that enables effective physiological tropism in inflamed plaque areas. Thus, reprogrammed immune cells capable of resolving inflammation may hold powerful potential in delivering effective therapies with fewer side effects often associated with pervasive small molecule inhibitors.

Recent studies defining the fundamental characteristics of immune cell dynamics are paving the way for the future development of immune cell therapeutics. Studies from our group and others reveal that innate leukocytes, including monocytes and neutrophils, can adopt dynamic polarization states depending upon the signal strength and duration of external danger/damage signals [32–35]. Persistent elevation of LDL-C and/or other risk factors, such as subclinical dosages of endotoxin, can synergistically reprogram monocytes or neutrophils into a chronic low-grade inflammatory state, characterized by elevated expression of adhesion molecules and reduced expression of homeostatic mediators, which collectively leads to their sustained infiltration and retention within the damaged vasculature leading to atherosclerosis pathogenesis [36–38]. Recent single cell–based trajectory analyses suggest that monocytes may first de-differentiate into a proliferative Ly6C high state [37, 39], which then further bifurcate into either an inflamed state with the sustained challenge of inflammatory signals [37] or remain at an anti-inflammatory proliferative state [39]. Our data are consistent with previous reports documenting the presence of a small subset of proliferative monocytes in vivo, which could exhibit either inflammatory or reparative properties [40–42].

Mechanistically, toll-like-receptor 4 (TLR4) serves as a key receptor or co-receptor for not only lipopolysaccharide (LPS) but also oxidized LDL (oxLDL), involved in the training of memory monocytes [43, 44]. We further defined key intracellular mechanisms responsive to the sustained generation of low-grade inflammatory monocyte memory [37]. There are two potentially competitive pathways, myeloid differentiating factor 88 (MyD88)-dependent and MyD88-independent pathways mediated by TRAM/TRIF downstream of TLR4 [45]. Past studies have mostly focused on the MyD88-pathway which is acute, robust yet transient, followed by negative feedback leading to anti-inflammatory tolerance [46, 47]. Relatively less-studied,

TRAM-dependent pathway appears to be persistent and long-lasting [48]. In sharp contrast to the acute and transient inflammatory effects induced by higher dose LPS, we first reported that pathologically relevant subclinical low-dose LPS potently induces persistent, non-resolving low-grade inflammatory responses of monocytes without tolerance [33, 36, 49]. We reported that TRAM mediates the generation of Ly6C^{hi} low-grade inflammatory monocytes [36, 50]. TRAM-deficient monocytes are resistant to low-dose LPS training and largely remain as Ly6C^{lo}-resolving monocytes [51, 52]. Consistent with our mechanistic findings, independent functional studies reported that myeloid deletion of TRAM, instead of MyD88, exhibits reduced pathogenesis of atherosclerosis [53].

Further, we found that TRAM signaling circuitry initiates the disruption of peroxisome homeostasis that is critical in sustaining an inflamed platform for the activation of SRC-related kinases and STAT1/STAT5 and the expression of inflammatory mediators [37]. The disruption of peroxisome homeostasis can lead to chronic elevation of cellular reactive oxygen species (ROS) which in turn sustains the activation of redox-sensitive kinases such as SRC, therefore forming a positive feedback loop in the generation of low-grade inflammatory memory monocytes [37, 54].

Our data further reveal that TRAM deletion enables the stable generation of "memory" resolving nonclassical Ly6C^{lo} monocytes with sustained expression of key anti-inflammatory resolving mediators (e.g., CD200R, catalase, ALOX15, and its resolvin product RvD1), due to elevated levels of PPARγ (the critical transcription factor involved in the expression of CD200R, ALOX15, and peroxisome genesis/pexophagy) and PEX5, correlated with enhanced pexophagy and reduced monocyte ROS [37]. We further demonstrated that resolving monocytes due to TRAM deletion can potently propagate inflammation resolution to neighboring leukocytes in vitro and in vivo via CD200R [37]. Our data are consistent with independent reports showing the anti-inflammatory efficacy of CD200R-mediated cellular communication [55–57]. The transfusion approach is becoming widely used to examine the in vivo efficacy of transfused cells and their interaction with recipient tissues given its less invasive nature and translational potential [58–61] as compared to bone marrow transplant which involves radiation-incurred extreme damage to the host immune environment and other compounding complications. Through fluorescent labeling, our studies and others indicate that transfused innate leukocytes can effectively reach host tissues, such as liver, lung, spleen, and aorta and remain alive for at least 5 days [38, 59–62]. When adoptively transfused into recipient mice, we found that TRAM-deficient monocytes can effectively propagate the resolving characteristics in vivo and reduce atherosclerosis pathogenesis [37]. Our data support the novel concept that enhancing pexophagy through TRAM deletion will enable the reprogramming of "resolving" monocytes capable of propagating inflammation resolution (Fig. 1).

In addition to the genetic manipulation of monocytes, we reported that chemical treatment of monocytes with 4-phenylbutyric acid (4-PBA) can similarly enable the generation of resolving monocytes [39]. 4-PBA is a potent peroxisome inducer and effectively activates PPAR-mediated transcription program of peroxisome synthesis, as well as anti-inflammatory mediators [51, 63, 64]. 4-PBA-trained monocytes retain both the proliferative and the anti-inflammatory potential which may hold promise in developing monocyte-based anti-atherosclerosis therapies.

Besides monocytes, other innate leukocytes such as neutrophils also play key modulatory roles during the pathogenesis and regression of atherosclerosis. Although mechanistically less studied, there is a compelling phenotypic connection between elevated circulating neutrophil counts with the development of atherosclerosis in both experimental animals and human patients [65–67]. Neutrophils account for ~50–70% of human circulating leukocytes. In human atherosclerosis, high circulating neutrophil ratios have been tightly associated with characteristics of rupture-prone atherosclerotic lesions [66, 68, 69]. Neutrophils were shown to be present within human atherosclerotic plaques [70]. In experimental animals, neutrophilia promoted by hyperlipidemia accelerates early atherosclerosis, which is apparent within 4 weeks of high-fat diet feeding [71]. Neutrophils contribute to atherosclerosis pathogenesis through enhanced adhesion to vasculatures and subsequent damage to endothelium by producing adhesion/swarming molecules, such as LTB4, ICAM-1, and tissue degrading molecules, such as Myeloperoxidase (MPO) and elastase [71–73]. Depletion of inflammatory neutrophils was shown to reduce the atherosclerotic lesion burden [71]. *Paradoxically*, neutrophils are also beneficial during the resolution of inflammation through expressing resolving mediators such as RvD1 [74, 75]. Recent clinical studies reveal that the ratios of LTB4 vs RvD1/RvE1 closely correlate with clinical severity of cardiovascular diseases in human patients [76, 77].

To harness the therapeutic potential of resolving neutrophils, we studied and reported that "resolving" neutrophils pharmacologically trained with 4-PBA exhibit enhanced peroxisome homeostasis capable of reducing atherosclerosis when transfused weekly into atherosclerotic mice [38]. 4-PBA-trained resolving neutrophils potently express elevated levels of resolving mediators, including RvD1, CD200R, and CD24, which can propagate homeostasis in recipient mice by improving vascular integrity and reducing tissue inflammation [38]. Genetically, we recently reported that TRAM deletion can also effectively reprogram neutrophils into a resolving state, with elevated homeostatic mediators including CD200R [78] (Fig. 2).

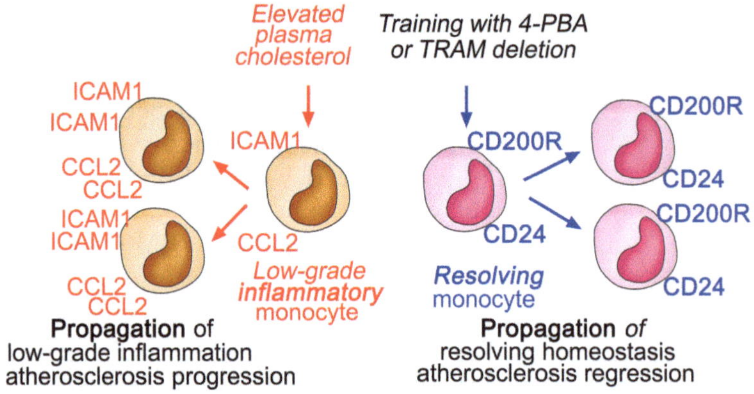

Fig. 1 Monocyte dynamics and therapeutic potential of resolving monocytes in treating atherosclerosis. Monocytes can be dynamically polarized into either a pro-atherogenic inflammatory state or an anti-atherogenic resolving state. Resolving monocytes with enhanced peroxisome homeostasis may propagate homeostasis and be amenable to the treatment of atherosclerosis

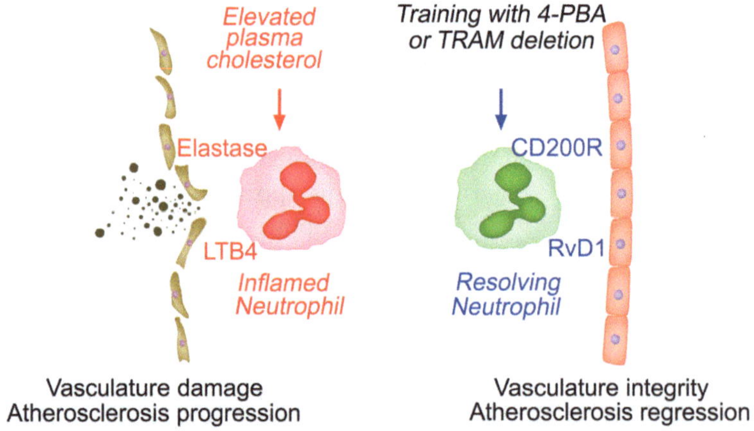

Fig. 2 Neutrophil dynamics and therapeutic potential of resolving neutrophils in treating atherosclerosis. Pro-inflammatory neutrophils play a key role during the initiation of vascular damage during the early stage of atherosclerosis, by secreting inflammatory mediators such as elastase and LTB4. On the other hand, certain subsets of neutrophils can suppress the degranulation and release of inflammatory mediators, while exhibiting enhanced expression of RvD1 and CD200R. Resolving neutrophils can be reprogrammed via genetic deletion of TRAM or chemical training with 4-PBA. Resolving neutrophils can maintain vasculature integrity

Since neutrophils can be naturally attracted to inflamed tissues, independent studies also explored using engineered neutrophils as a natural carrier to deliver anti-inflammatory drugs to achieve more effective tissue delivery to treat atherosclerosis [79]. Similar approaches have been developed to deliver anti-cancer drugs loaded onto neutrophils [80]. It was shown that neutrophils can be readily guided into the inflamed brain tissues by inflammatory mediators released following tumor resection. Neutrophils engineered to carry paclitaxel can then readily penetrate the blood-brain barrier and effectively reduce glioma growth [80]. Independent studies reported that vascular inflammation can be reduced by targeting neutrophils with anti-inflammatory nanoparticles [81, 82].

Besides intact innate leukocytes, monocyte- or neutrophil-derived extracellular vehicles (EVs) have also been developed to deliver anti-inflammatory therapeutics. EVs generated from disrupted neutrophils carry less subcellular organelles and have shown to be effective in serving as a drug delivery platform to treat inflammatory diseases [83].

4 Challenges and Future Outlooks

Despite significant advancements in the emerging areas of immune cell–based therapeutics during the last decades, significant hurdles still remain. The most pressing issue is still the lack of a clear characterization of complex innate immune cell polarization dynamics. Emerging studies continuously reveal increasing amounts of monocyte and neutrophil subsets, with unique pro- or anti-inflammatory features. Although recent studies powered by single cell–sequencing approaches have enabled better profiling of gene expressions from diverse subsets, functional validation and clarification are still lacking. It is also not clear how stable or transient these subsets of innate leukocytes would remain within the inflamed environment. The other less-defined issue is in regard to the constant inter-cellular communication between innate leukocytes and neighboring cells. The complex kinetics and dynamics of immune cell interactions within the local inflammatory niche still warrant extensive studies. Nevertheless, future single cell–based real-time studies with spatial resolution will better enable the characterization of engineered immune cells in the treatment of atherosclerosis.

5 Conclusion

In perspective, we anticipate better characterization and efficient isolation of leukocyte subsets with well-defined resolving potentials in the future, which will enable targeted treatment of inflamed atherosclerotic tissues. Complementary to the cell-based approach, we also anticipate improved delivery of molecule-based therapeutics with engineered leukocytes or leukocyte-based extracellular vesicles. Integrated efforts of medicinal chemistry and immune-based delivery will continue driving the effective development of next-generation therapeutics to fill the unmet medical needs for the effective management of cardiovascular disease.

Acknowledgements

We would like to thank Grace Lee, who generated the figure illustrations for this manuscript.

Funding

Dr. Li is supported by grants from NIH R01HL115835.

References

1. Herrington W, Lacey B, Sherliker P et al (2016) Epidemiology of atherosclerosis and the potential to reduce the global burden of atherothrombotic disease. Circ Res 118(4): 535–546
2. Vaduganathan M, Mensah GA, Turco JV et al (2022) The global burden of cardiovascular diseases and risk: a compass for future health. J Am Coll Cardiol 80(25):2361–2371
3. Abe RJ, Abe JI, Nguyen MTH et al (2022) Free cholesterol bioavailability and atherosclerosis. Curr Atheroscler Rep 24(5):323–336
4. Hedayatnia M, Asadi Z, Zare-Feyzabadi R et al (2020) Dyslipidemia and cardiovascular disease risk among the MASHAD study population. Lipids Health Dis 19(1):42
5. Silverman MG, Ference BA, Im K et al (2016) Association between lowering LDL-C and cardiovascular risk reduction among different therapeutic interventions: a systematic review and meta-analysis. JAMA 316(12):1289–1297
6. Galkina E, Ley K (2009) Immune and inflammatory mechanisms of atherosclerosis (*). Annu Rev Immunol 27:165–197
7. Soehnlein O, Libby P (2021) Targeting inflammation in atherosclerosis – from experimental insights to the clinic. Nat Rev Drug Discov 20(8):589–610
8. Ridker PM, Everett BM, Thuren T et al (2017) Antiinflammatory therapy with canakinumab for atherosclerotic disease. N Engl J Med 377(12):1119–1131
9. Ricciotti E, FitzGerald GA (2011) Prostaglandins and inflammation. Arterioscler Thromb Vasc Biol 31(5):986–1000
10. Force USPST, Davidson KW, Barry MJ et al (2022) Aspirin use to prevent cardiovascular disease: US preventive services task force recommendation statement. JAMA 327(16): 1577–1584
11. Liu Y, Yang F, Zou S et al (2018) Rapamycin: a bacteria-derived immunosuppressant that has anti-atherosclerotic effects and its clinical application. Front Pharmacol 9:1520
12. Weichhart T, Hengstschlager M, Linke M (2015) Regulation of innate immune cell function by mTOR. Nat Rev Immunol 15(10): 599–614
13. Byles V, Covarrubias AJ, Ben-Sahra I et al (2013) The TSC-mTOR pathway regulates macrophage polarization. Nat Commun 4: 2834
14. Boada C, Zinger A, Tsao C et al (2020) Rapamycin-loaded biomimetic nanoparticles reverse vascular inflammation. Circ Res 126(1):25–37
15. Kurdi A, De Meyer GR, Martinet W (2016) Potential therapeutic effects of mTOR inhibition in atherosclerosis. Br J Clin Pharmacol 82(5):1267–1279
16. Liao JK (2005) Effects of statins on 3-hydroxy-3-methylglutaryl coenzyme a reductase inhibition beyond low-density lipoprotein cholesterol. Am J Cardiol 96(5A):24F–33F
17. Sacks FM, Pfeffer MA, Moye LA et al (1996) The effect of pravastatin on coronary events after myocardial infarction in patients with average cholesterol levels. Cholesterol and Recurrent Events Trial investigators. N Engl J Med 335(14):1001–1009
18. Kwon O, Kang SJ, Kang SH et al (2017) Relationship between serum inflammatory marker levels and the dynamic changes in coronary plaque characteristics after statin therapy. Circ Cardiovasc Imaging 10(7):e005934

19. Ward NC, Watts GF, Eckel RH (2019) Statin toxicity. Circ Res 124(2):328–350
20. Rashid S, Tavori H, Brown PE et al (2014) Proprotein convertase subtilisin kexin type 9 promotes intestinal overproduction of triglyceride-rich apolipoprotein B lipoproteins through both low-density lipoprotein receptor-dependent and -independent mechanisms. Circulation 130(5):431–441
21. Shimada YJ, Cannon CP (2015) PCSK9 (Proprotein convertase subtilisin/kexin type 9) inhibitors: past, present, and the future. Eur Heart J 36(36):2415–2424
22. Lagace TA (2014) PCSK9 and LDLR degradation: regulatory mechanisms in circulation and in cells. Curr Opin Lipidol 25(5):387–393
23. Abifadel M, Varret M, Rabes JP et al (2003) Mutations in PCSK9 cause autosomal dominant hypercholesterolemia. Nat Genet 34(2): 154–156
24. Doi T, Hori M, Harada-Shiba M et al (2021) Patients with LDLR and PCSK9 gene variants experienced higher incidence of cardiovascular outcomes in heterozygous familial hypercholesterolemia. J Am Heart Assoc 10(4):e018263
25. Punch E, Klein J, Diaba-Nuhoho P et al (2022) Effects of PCSK9 targeting: alleviating oxidation, inflammation, and atherosclerosis. J Am Heart Assoc 11(3):e023328
26. Barale C, Melchionda E, Morotti A et al (2021) PCSK9 biology and its role in atherothrombosis. Int J Mol Sci 22(11):5880
27. Kong P, Cui ZY, Huang XF et al (2022) Inflammation and atherosclerosis: signaling pathways and therapeutic intervention. Signal Transduct Target Ther 7(1):131
28. Hampton EN, Knuth MW, Li J et al (2007) The self-inhibited structure of full-length PCSK9 at 1.9 A reveals structural homology with resistin within the C-terminal domain. Proc Natl Acad Sci U S A 104(37): 14604–14609
29. Chaudhary R, Garg J, Shah N et al (2017) PCSK9 inhibitors: a new era of lipid lowering therapy. World J Cardiol 9(2):76–91
30. Zhang XL, Zhu QQ, Zhu L et al (2015) Safety and efficacy of anti-PCSK9 antibodies: a meta-analysis of 25 randomized, controlled trials. BMC Med 13:123
31. Gurgoze MT, Muller-Hansma AHG, Schreuder MM et al (2019) Adverse events associated with PCSK9 inhibitors: a real-world experience. Clin Pharmacol Ther 105(2):496–504
32. Geng S, Pradhan K, Li L (2022) Signal-strength and history-dependent innate immune memory dynamics in health and disease. Handb Exp Pharmacol 276:23–41
33. Yuan R, Geng S, Li L (2016) Molecular mechanisms that underlie the dynamic adaptation of innate monocyte memory to varying stimulant strength of TLR ligands. Front Immunol 7:497
34. Chen K, Geng S, Yuan R et al (2015) Super-low dose endotoxin pre-conditioning exacerbates sepsis mortality. EBioMedicine 2(4): 324–333
35. Lee C, Geng S, Zhang Y et al (2017) Programming and memory dynamics of innate leukocytes during tissue homeostasis and inflammation. J Leukoc Biol 102(3):719–726
36. Geng S, Chen K, Yuan R et al (2016) The persistence of low-grade inflammatory monocytes contributes to aggravated atherosclerosis. Nat Commun 7:13436
37. Geng S, Zhang Y, Yi Z et al (2021) Resolving monocytes generated through TRAM deletion attenuate atherosclerosis. JCI Insight 6(20): e149651
38. Geng S, Zhang Y, Lee C et al (2019) Novel reprogramming of neutrophils modulates inflammation resolution during atherosclerosis. Sci Adv 5(2):eaav2309
39. Lee J, Geng S, Li S et al (2021) Single cell RNA-seq and machine learning reveal novel subpopulations in low-grade inflammatory monocytes with unique regulatory circuits. Front Immunol 12:627036
40. Swirski FK, Hilgendorf I, Robbins CS (2014) From proliferation to proliferation: monocyte lineage comes full circle. Semin Immunopathol 36(2):137–148
41. Lin JD, Nishi H, Poles J et al (2019) Single-cell analysis of fate-mapped macrophages reveals heterogeneity, including stem-like properties, during atherosclerosis progression and regression. JCI Insight 4(4):e124574
42. Hardtner C, Kornemann J, Krebs K et al (2020) Inhibition of macrophage proliferation dominates plaque regression in response to cholesterol lowering. Basic Res Cardiol 115(6):78
43. Wiesner P, Choi SH, Almazan F et al (2010) Low doses of lipopolysaccharide and minimally oxidized low-density lipoprotein cooperatively activate macrophages via nuclear factor kappab and activator protein-1: possible mechanism for acceleration of atherosclerosis by subclinical endotoxemia. Circ Res 107(1):56–65
44. Bae YS, Lee JH, Choi SH et al (2009) Macrophages generate reactive oxygen species in response to minimally oxidized low-density lipoprotein: toll-like receptor 4- and spleen tyrosine kinase-dependent activation of

NADPH oxidase 2. Circ Res 104(2):210–218. 221p following 218
45. Kawai T, Akira S (2007) TLR signaling. Semin Immunol 19(1):24–32
46. Morris MC, Gilliam EA, Li L (2014) Innate immune programming by endotoxin and its pathological consequences. Front Immunol 5: 680
47. Piao W, Song C, Chen H et al (2009) Endotoxin tolerance dysregulates MyD88- and toll/IL-1R domain-containing adapter inducing IFN-beta-dependent pathways and increases expression of negative regulators of TLR signaling. J Leukoc Biol 86(4):863–875
48. Cheng Z, Taylor B, Ourthiague DR et al (2015) Distinct single-cell signaling characteristics are conferred by the MyD88 and TRIF pathways during TLR4 activation. Sci Signal 8(385):ra69
49. Yuan R, Geng S, Chen K et al (2016) Low-grade inflammatory polarization of monocytes impairs wound healing. J Pathol 238(4):571–583
50. Rahtes A, Li L (2020) Polarization of low-grade inflammatory monocytes through TRAM-mediated up-regulation of keap1 by super-low dose endotoxin. Front Immunol 11:1478
51. Rahtes A, Pradhan K, Sarma M et al (2020) Phenylbutyrate facilitates homeostasis of non-resolving inflammatory macrophages. Innate Immun 26(1):62–72
52. Pradhan K, Geng S, Zhang Y et al (2021) TRAM-related TLR4 pathway antagonized by Irak-M mediates the expression of adhesion/coactivating molecules on low-grade inflammatory monocytes. J Immunol 206(12): 2980–2988
53. Lundberg AM, Ketelhuth DF, Johansson ME et al (2013) Toll-like receptor 3 and 4 signalling through the TRIF and TRAM adaptors in haematopoietic cells promotes atherosclerosis. Cardiovasc Res 99(2):364–373
54. Forrester SJ, Kikuchi DS, Hernandes MS et al (2018) Reactive oxygen species in metabolic and inflammatory signaling. Circ Res 122(6): 877–902
55. van der Vlist M, Raoof R, Meyaard L, Eijkelkamp N (2022) Macrophages transfer mitochondria to sensory neurons to resolve inflammatory pain. Neuron 110(4):613–626. e9
56. Ngwa C, Liu F (2019) CD200-CD200R signaling and diseases: a potential therapeutic target? Int J Physiol Pathophysiol Pharmacol 11(6):297–309
57. Kassiteridi C, Cole JE, Griseri T et al (2021) CD200 limits monopoiesis and monocyte recruitment in atherosclerosis. Circ Res 129(2):280–295
58. Nemeth T, Sperandio M, Mocsai A (2020) Neutrophils as emerging therapeutic targets. Nat Rev Drug Discov 19(4):253–275
59. Zhang Y, Lee C, Geng S et al (2020) Neutrophils deficient in innate suppressor IRAK-M enhances anti-tumor immune responses. Mol Ther 28(1):89–99
60. Zhang Y, Lee C, Geng S et al (2019) Enhanced tumor immune surveillance through neutrophil reprogramming due to Tollip deficiency. JCI Insight 4(2):e122939
61. Zhang Y, Lin R, Pradhan K et al (2020) Innate priming of neutrophils potentiates systemic multiorgan injury. Immunohorizons 4(7): 392–401
62. Ran T, Zhang Y, Diao N et al (2019) Enhanced neutrophil immune homeostasis due to deletion of PHLPP. Front Immunol 10:2127
63. Liu N, Qiang W, Kuang X et al (2002) The peroxisome proliferator phenylbutyric acid (PBA) protects astrocytes from ts1 MoMuLV-induced oxidative cell death. J Neurovirol 8(4): 318–325
64. Di Cara F, Andreoletti P, Trompier D et al (2019) Peroxisomes in immune response and inflammation. Int J Mol Sci 20(16):3877
65. Silvestre-Roig C, Braster Q, Ortega-Gomez A et al (2020) Neutrophils as regulators of cardiovascular inflammation. Nat Rev Cardiol 17(6):327–340
66. Adamstein NH, MacFadyen JG, Rose LM et al (2021) The neutrophil-lymphocyte ratio and incident atherosclerotic events: analyses from five contemporary randomized trials. Eur Heart J 42(9):896–903
67. Mawhin MA, Tilly P, Zirka G et al (2018) Neutrophils recruited by leukotriene B4 induce features of plaque destabilization during endotoxaemia. Cardiovasc Res 114(12):1656–1666
68. Nam KW, Kwon HM, Jeong HY et al (2018) High neutrophil to lymphocyte ratios predict intracranial atherosclerosis in a healthy population. Atherosclerosis 269:117–121
69. Ionita MG, van den Borne P, Catanzariti LM et al (2010) High neutrophil numbers in human carotid atherosclerotic plaques are associated with characteristics of rupture-prone lesions. Arterioscler Thromb Vasc Biol 30(9): 1842–1848
70. Fernandez DM, Rahman AH, Fernandez NF et al (2019) Single-cell immune landscape of human atherosclerotic plaques. Nat Med 25(10):1576–1588

71. Drechsler M, Megens RT, van Zandvoort M et al (2010) Hyperlipidemia-triggered neutrophilia promotes early atherosclerosis. Circulation 122(18):1837–1845
72. Rotzius P, Thams S, Soehnlein O et al (2010) Distinct infiltration of neutrophils in lesion shoulders in ApoE-/- mice. Am J Pathol 177(1):493–500
73. Li J, Kumari T, Barazia A et al (2022) Neutrophil DREAM promotes neutrophil recruitment in vascular inflammation. J Exp Med 219(1): e20211083
74. Casanova-Acebes M, Nicolas-Avila JA, Li JL et al (2018) Neutrophils instruct homeostatic and pathological states in naive tissues. J Exp Med 215(11):2778–2795
75. Filep JG, Ariel A (2020) Neutrophil heterogeneity and fate in inflamed tissues: implications for the resolution of inflammation. Am J Physiol Cell Physiol 319(3):C510–C532
76. Thul S, Labat C, Temmar M et al (2017) Low salivary resolvin D1 to leukotriene B4 ratio predicts carotid intima media thickness: a novel biomarker of non-resolving vascular inflammation. Eur J Prev Cardiol 24(9): 903–906
77. Welty FK, Schulte F, Alfaddagh A et al (2021) Regression of human coronary artery plaque is associated with a high ratio of (18-hydroxy-eicosapentaenoic acid + resolvin E1) to leukotriene B4. FASEB J 35(4):e21448
78. Lin R, Yi Z, Wang J et al (2022) Generation of resolving memory neutrophils through pharmacological training with 4-PBA or genetic deletion of TRAM. Cell Death Dis 13(4):345
79. Xue Y, Wu Y, Wang Q et al (2019) Cellular vehicles based on neutrophils enable targeting of atherosclerosis. Mol Pharm 16(7): 3109–3120
80. Xue J, Zhao Z, Zhang L et al (2017) Neutrophil-mediated anticancer drug delivery for suppression of postoperative malignant glioma recurrence. Nat Nanotechnol 12(7): 692–700
81. Wang Z, Li J, Cho J et al (2014) Prevention of vascular inflammation by nanoparticle targeting of adherent neutrophils. Nat Nanotechnol 9(3):204–210
82. Chu D, Gao J, Wang Z (2015) Neutrophil-mediated delivery of therapeutic nanoparticles across blood vessel barrier for treatment of inflammation and infection. ACS Nano 9(12): 11800–11811
83. Gao J, Wang S, Wang Z (2017) High yield, scalable and remotely drug-loaded neutrophil-derived extracellular vesicles (EVs) for anti-inflammation therapy. Biomaterials 135:62–73

Chapter 3

T Cell Subsets and Immune Homeostasis

Chuyu Wu, Mei-Ling Jiang, Tao Pang, and Cun-Jin Zhang

Abstract

T cells are a heterogeneous group of cells that can be classified into different subtypes according to different classification methods. The body's immune system has a highly complex and effective regulatory network that allows for the relative stability of immune system function. Maintaining proper T cell homeostasis is essential for promoting protective immunity and limiting autoimmunity and tumor formation. Among the T cell family members, more and more T cell subsets have gradually been characterized. In this chapter, we summarize the functions of some key T cell subsets and their impact on immune homeostasis.

Key words T cells, T cell subset, Immune homeostasis, Helper T cells

1 Introduction

T cells (T lymphocytes, thymus-dependent lymphocytes) are the major component of lymphocytes that play an important role in immune response and function by secreting soluble cytokines or through cellular contact to defend against disease infection and tumor formation in the body [1]. Early in embryonic development, T cell precursor stem cells are transported to the thymus via the bloodstream, migrate from the superficial cortex to the deep cortex and medulla, and develop into mature T cells (Fig. 1). After receiving orderly and standardized "training," mature T cells enter the bloodstream and are transferred to peripheral lymphoid tissues (e.g., spleen, lymph nodes, etc.), where they are stimulated and then differentiated into effector or memory T cells that participate in adaptive immunity [1–3].

The precise localization of T cells is related to their expression of homing receptors, such as L-selectin and sphingosine 1-phosphate1 (S1P1) protein [4]. Upon contact with antigens, naïve T cells differentiate and develop into different functional T cell subpopulations and express new homing receptors, thus "homing" to new sites [4, 5]. The classification of T cells is based on a

Vijay Kumar (ed.), *Immune Homeostasis: Methods and Protocols*, Methods in Molecular Biology, vol. 2782,
https://doi.org/10.1007/978-1-0716-3754-8_3,

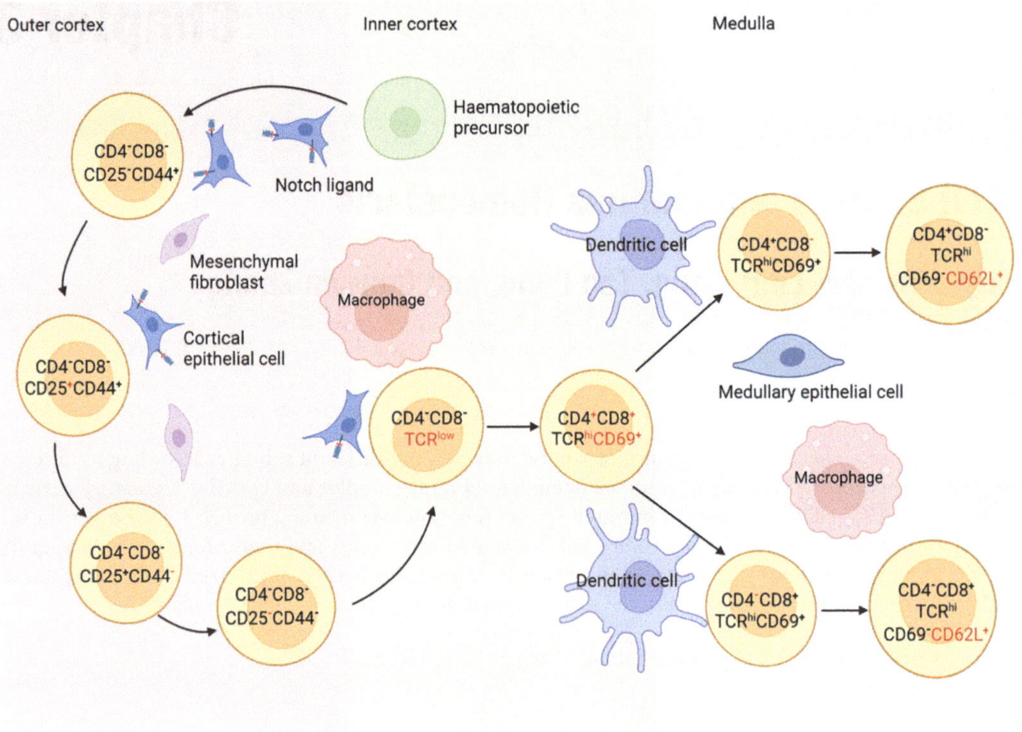

Fig. 1 Developmental process of T cells. T cell precursor stem cells are transported to the thymus through the bloodstream and migrate from the superficial cortex to the deep cortex and medulla. T cell precursor stem cells in the thymus undergo TCR rearrangement → positive selection → negative selection → maturation of naïve $CD4^+$/$CD8^+$ T cells

variety of bases, and the different types in the peripheral and tumor microenvironments add to the difficulty of understanding them. Cell surface molecules and intracellular transcription factors are the basis for T cell sorting and are artificial markers to facilitate research and communication (Table 1). CD45 molecules are single-chain transmembrane proteins that are expressed in all immune cells [6, 7]. Different subsets of T cells express different CD45 molecular isoforms and are often used to distinguish the differentiation status of T cells. The T cell receptor (TCR)-CD3 complex is a molecular marker specific to T cells and is expressed on the surface of all T cells [8]. The TCR recognizes specific antigens, and CD3 transmits signals to the cell, which together perform signaling functions [8].

The immune system is a tightly regulated network capable of maintaining homeostasis of immune homeostasis under normal physiological conditions [9]. Normally, when attacked by foreign antigens, specific appropriate responses aimed at restoring homeostasis in the body are initiated. However, when immune homeostasis is disrupted under specific circumstances, this balance cannot be

Table 1
T cell classification

T cell subset	Surface markers	Transcription factor	Function
γδ T cell			
γδ T cell	γδ, TCR, CD3	IFN-Y, IL-17A, IL-17F, IL-22	γδ T cell is often enriched on the surface of epithelial cells, which can promote inflammation or anti-inflammatory, depending on the microenvironment. It has the characteristics of both innate immunity and adaptive immunity.
Naïve T cell			
CD4+ αβ T cell	αβTCR, CD3, CD4, CCR7, $CD62L^{hi}$, IL-7R (CD127)	THPOK	Patrol and monitor MHC II peptide molecular complexes in lymph nodes. Upon activation by APCs, naïve CD4+ T cells differentiate into effector T cells or regulatory T cells. Activated naïve T cells also give rise to memory T cells.
CD8+ αβ T cell	αβTCR, CD3, CD8, CCR7, $CD62L^{hi}$, IL-7R (CD127)	RUNX3	Patrol and monitor the MHC1 peptide molecular complex in lymph nodes. The naïve T cells differentiated into cytotoxic T cells and memory T cells after being activated by APCs.
Effector T cell (Cytotoxic T lymphocytes, helper T cell, regulatory T cell, natural killer T cell)			
CTL cell	αβTCR, CD3, CD8	EOMES, T-bet, BLIMP1	CTL has cytotoxicity, which protects the host from virus infection and cancer by killing infected and mutated cells. CTL can also directly kill target cells by secreting perforin and granzyme, as well as induce apoptosis in target cells.
Th1 cell	αβTCR, CD3, CD4, IL-12R, IFNγR, CXCR3	T-bet, STAT4, STAT1	Effector molecule: IFNγ, IL-2, LTα Promote the protective immunity for intracellular pathogens. By secreting IFNγ, activating macrophages, and raising INOS, it can kill intracellular pathogenic bacteria. The development of Th1 cell is regulated by IL-12.

(continued)

Table 1 (continued)

T cell subset	Surface markers	Transcription factor	Function
Th2 cell	αβTCR, CD3, CD4, IL-4R, IL-33R, CCR4, IL-17RB, CRTH2	GATA3, STAT6, DEC2, MAF	Effector molecule: IL-4, IL-5, IL-13, IL-10 Promote humoral immunity and host response against extracellular pathogens. Promote the occurrence of allergy and asthma. The development of Th2 cell is regulated by IL-4, IL-25, and IL-33.
Th9 cell	αβTCR, CD3, CD4	PU.1	Effector molecule: IL-9, IL-10 Participate in the host to extracellular parasites. Promote allergic inflammation.
Th17 cell	αβTCR, CD3, CD4, IL-23R, CCR6, IL-1R	RORγt, STAT3, RORα	Effector molecule: IL-17A, IL-17F, IL-21, IL-22, CCL20 Promote protective immunity against extracellular bacteria and fungi. Induce autoimmune diseases and inflammation. Th17 cells are produced in the presence of TGF β and IL-6/IL-21 and are dependent on IL-23 and IL-1 for maintenance.
Th22 cell	αβTCR, CD3, CD4, CCR10	AHR	Effector molecule: IL-22 Identified in inflammatory dermatosis. The function is unknown.
Tfh cell	αβTCR, CD3, CD4, CXCR5, SLAM, OX40L, CD40L, ICOS, IL-21R, PD1	BCL-6, STAT3	Effector molecule: IL-21 Participate in the reaction of germinal center and provide auxiliary function for B cell-type conversion.
nTreg cell	αβTCR, CD3, CD4, CD25, CTLA4, GITR	FOXP3, STAT5, FOXO1, FOXO3	Effector molecule: IL-10, TGFβ, IL-35 Mediate immunosuppression and immune tolerance through contact or noncontact mechanism.
iTreg cell	αβTCR, CD3, CD4, CD25, CTLA4, GITR	FOXP3, FOXO1, FOXO3, STAT5, SMAD2, SMAD3, SMAD4	Effector molecule: IL-10, TGFβ Mediate immunosuppression and immune tolerance through contact or noncontact mechanism.

(continued)

Table 1
(continued)

T cell subset	Surface markers	Transcription factor	Function
T_R1 cell	αβ TCR, CD3, CD4	–	Effector molecule: IL-10 T_R1 cell exerts its immunosuppressive function by producing IL-10. The T_R1 cells are differentiated from the naïve T cells under the conditions of VD3 and dexamethasone or TGFβ and IL-27.
NKT cell	NK1.1, SLAMF1, SLAMF6, TGFβR	PLZF	Effector molecule: IL-4, IFNγ, IL-17A NKT cell can promote inflammation as well as anti-inflammatory. It plays an immune regulatory role in cancer, autoimmune diseases, allergies, infections, GVHD, and other diseases.
Memory T cell			
Central memory T cell (Tcm cell)	$CCR7^{hi}$, CD44, $CD62L^{hi}$, TCR, CD3, IL-7R (CD127), IL-15R	BCL-6, BCL-6B, MBD2, BMI1	It preferentially resides in secondary lymphoid organs to increase re-stimulation response to antigen. Once stimulated by antigen, it rapidly proliferates and differentiates into effector T cells.
Effect memory T cell (Tem cell)	$CD62L^{low,}$ CD44, TCR, CD3, IL-7R (CD127), IL-15R, $CCR7^{low}$	BLIMP1	Once stimulated by antigen, it can rapidly produce effector cytokines and provide immune protection.
Other types of T cells			
Anergic T cell	CD3, CD8, PD1, TIM3, 1B11, LAG3	GRAIL, CBL-B, ITCH, NEDD4	Anergic T cell belongs to the cell with inactive function and cannot proliferate or produce IL-2. It may be important to avoid autoimmune reactions.
Exhausted T cell	CD3, CD8, PD1, TIM3, 1B11, LAG3	BLIMP1	As antigen-mediated TCR suffers from long-term chronic stimulation and the cells express more inhibitory receptors and lack the ability to produce effector cytokines, exhausted T cells cannot play an effective anti-viral role.

maintained, resulting in an inadequate or overreactive immune response. Cancer is an example of how the immune response may be inefficient or unresponsive, leading to uncontrolled growth of cancer cells. Conversely, when the immune response is excessive, this can lead to autoimmune disease. Immune homeostasis ensures that the number, diversity, and function of lymphocytes are maintained. This seemingly "quiescent" condition is the result of mechanisms that actively maintain various leukocyte populations and keep them functionally specialized. When functioning properly, homeostatic mechanisms serve to maintain an appropriate balance between effector and regulatory cells, preventing unwanted immunopathology and autoimmune disease. Therapeutic manipulation of immune homeostasis can be used to restore this balance in patients with chronic infections or autoimmune diseases and to prevent graft rejection or to promote immune responses against tumors.

2 γδ T Cells

Gamma delta T cells (γδ T cells) are one of the earliest developed T cells in the thymus, and their surface TCRs consist of γ and δ chains [10], accounting for about 1–5% of the number of T cells in the body, mainly in epithelial and mucosal tissues, such as small intestine, skin, lung, etc. [11]. γδ T cells have a natural homing advantage. γδ TCRs do not need to bind to major histocompatibility complex (MHC) molecules, do not need antigen-presenting cells (APCs) to present, and can directly recognize and bind antigen molecules [10].

Human γδ T cells can be divided into four major populations (δ1, δ2, δ3, and δ5) based on TCRδ chain expression [10]. Vδ9 T cells co-expressing the Vγ2 chain are the most common subpopulation of the γδ T cells population in the peripheral blood and secondary lymphoid organs of fetal and adult individuals [12–14]. Although Vγ9Vδ2 T cells are primarily localized in the circulation, they can be recruited to inflamed tissues and participate in pathogen and tumor clearance; however, they also promote inflammation. In many microbial infections, γδ T cells increase locally and/or systemically a few days after infection and are able to reach 50% of the total number of circulating T cells [15]. The rapid effector responses elicited during infection are similar to those produced by innate immune cells, a property associated with their ability to be activated in the absence of antigen initiation [16]. γδ T cells do so by releasing cytotoxic granules and inhibitory or lytic molecules such as granulysin and defensins [17]. Moreover, they have an indirect role in the elimination of microorganisms by producing pro-inflammatory cytokines and inducing antimicrobial functions in immune cells and epithelial cells.

It is well known that IL-17A plays an important role in the progression of autoimmune diseases [18]. Although the main source of IL-17A is the Th17CD4^{+}αβ T cell population, γδ T cells also contribute to IL-17A at the onset of autoimmune disease [19]. Human IL-17A-producing γδ T cells are produced in the periphery and can be recruited to inflammatory tissues, where they accumulate [20]. This process occurs more rapidly than the activation of conventional T lymphocytes. This is because γδ T cells can be activated in the absence of homologous TCR ligands, which makes them powerful early inflammatory inducers in autoimmune diseases.

The role of γδ T cells in psoriasis and multiple sclerosis has been extensively studied. Large numbers of γδ T cells are present in skin biopsies from patients with psoriasis. In patients with multiple sclerosis, γδ T cells accumulate in plaques and areas of chronic demyelination of the central nervous system (CNS). And in the cerebrospinal fluid, IL-17A-producing γδ T cells were increased [21], and IL-17 was elevated in peripheral blood and CNS [22]. In addition, γδ T cells have potent antitumor properties and can kill tumor cells either directly or indirectly. γδ T cells directly kill tumor cells by releasing cytotoxic substances such as perforin, granzyme, and cytokines (IFN-γ and TNF-α) after migration to the local tumor environment by recognizing NK cell receptors on the cell surface [23, 24]. γδ T cells can also synergistically activate B/DC (dentritic cell)/αβ T/NK (natural killer) cells. For example, they can act as APCs to activate αβ T cells or induce NK-mediated antitumor cytotoxicity by the 4-1BB co-stimulatory pathway [25], thus achieving indirect tumor killing.

3 Naïve T Cell

Naïve T cells are divided into two main categories: CD4^{+} αβ and CD8^{+} αβ, whose main function is to patrol lymph nodes and monitor MHC molecules [26]. Once activated by APCs, the former can rapidly differentiate into effector T cells or regulatory T cells, while the latter can differentiate into cytotoxic T cells or memory T cells.

Even though millions of naïve T cells are produced from the thymus every day, the number of T cells in the periphery is almost constant [26]. This homeostasis may require strict regulation. Survival signals for naïve T cells include TCR-pMHC interactions and cytokines such as IL-7. IL-7 has the dominant role in in vivo T cell survival [27, 28]. IL-7 is produced at constitutive levels by stromal cells resident in various organs, as well as by thymic and intestinal epithelial cells. Under physiologic conditions, IL-7 supports the long-term survival of naïve and memory T cells without inducing proliferation, thereby maintaining the regular size of the T cell pool

[29]. Under conditions of lymphopenia, the concentration of IL-7 rises to supra homeostatic levels that induce proliferation of both naïve and memory T cells with the aim of reconstituting the physiologic T cell pool, a process commonly referred to as lymphopenia-induced proliferation. The survival of naïve T cells is impaired by injecting mice with monoclonal IL-7-specific blocking antibodies [30] or after transfer into IL-7-deficient mice [31].

4 Memory T Cells

Memory T cells have encountered and react to a specific antigens and can recognize foreign invaders such as bacteria, viruses, and cancer cells [32]. Memory T cells are usually formed as a result of having been infected with a specific antigen, having experienced cancer, or having been vaccinated. When the same antigen enters the body again, memory T cells can generate a more intense immune response more quickly than when first exposed to this antigen [32]. Memory T cells are divided into central memory T cells (Tcm) and effector memory T cells (Tem) [33]. Tcm express CCR7 and CD62L proteins and secrete interleukin-2 but do not produce IFN-γ and IL-4. Tem cells do not express CCR7 and CD62L proteins but produce two cytokines such as IFN-γ, IL-4, IL-5, and perforin.

Recently, it has been discovered that most memory cells in non-lymphoid tissues are not actually Tem cells, but another type of memory cell. These cells are permanently present in the tissue and are therefore called tissue-resident memory T cells (Trm) [34]. In some cases, Trm can elicit rapid in situ protective responses. Trm form part of the immune sensing network, have the ability to control local immune homeostasis, and are involved in immune responses mediated by pathogens, cancers, and possibly autoantigens during autoimmunity. Mouse $CD4^+$ Trm cells can be generated in the lung by the relay transfer of activated (effector) T cells or after infection with respiratory viruses [35], and they have an enhanced ability to mediate protection against influenza virus infection compared to circulating memory $CD4^+$ T cells [36]. A similar subpopulation of noncirculating $CD4^+$ TRM cells has been identified in the bone marrow of mice after systemic viral infection, exhibiting enhanced paracrine function [37].

The number of circulating memory T lymphocytes with different immune responses decreases slowly over time. However, cells can persist through steady-state proliferation or as resting cells independent of antigen [38]. Metastatic splenic memory T lymphocytes are home to a variety of organs, including lung, liver, blood, and bone marrow, and within 25 days, 30%–60% of them have proliferated at least once in the organs analyzed [39]. Memory T cells have a half-life of less than 2 weeks, and 50% of memory-proliferating T cells divide at least once within 2 weeks. This proliferation has been shown to be dependent on cytokines IL-7 and IL-15 [40, 41]. IL-7 is required for memory $CD4^+$ cells, and memory $CD8^+$ T cells are dependent on IL-15 [42, 43].

5 Effector T Cells

Effector T cells are cells formed by the proliferation and differentiation of T cells after receiving antigenic stimulation. Effector T cells describe a group of cells that include several T cell types that actively respond to stimulation (e.g., co-stimulation), including helper T cells, cytotoxic T cells, and regulatory T cells, and effector T cells have the function of releasing lymphokines [44].

Cytotoxic T Lymphocyte The main surface marker of cytotoxic T cells (CTL) is CD8, also known as killer T cells. The main function of CTL is to secrete perforins and granzymes that kill infected and mutated cells [45]. Typically, $CD8^+$ T cells are studied as a whole, but they can equally be further divided into different subtypes. As with helper T cells, each subset can be defined by surface proteins, transcription factors, and cytokines produced. Subsets and associated markers include [46] Tcl (TNF-α, INF-γ, IL-2, CXCR3, TBX21); Tc2 (IL-4, IL-5, CCR4, GATA3); Tc9 (IL-9, IL-10, IRF4); Tc17 (CCR6, KLRB1, IL-17, IRF4, RORC); and Tc22 (IL-2, IL-22, TNF-α).

The two mechanisms of $CD8^+$ T cell killing of target cells are the granzyme [47, 48] pathway and the Fas pathway [49, 50]. T cell receptors interact with antigens presented on the surface of target cells to form aggregates. The Golgi apparatus in CTL starts to accumulate, and granulase-containing granules re-localize to the area in contact with the target cells and release perforin and granzyme through cytokinesis, which enters the target cells and activates their apoptotic pathway, inducing cell death [47]. Some $CD8^+$ T cells lack perforin and granzyme, and they mediate cytotoxicity via the Fas pathway. Fas ligands (Fas-L, CD178) on the cell surface bind to Fas (CD95) on target cells, causing them to bind Fas-associated death domain proteins, which accordingly trigger a series of responses leading to apoptosis.

6 T Helper Cells

T helper cells (TH) have the function of assisting in humoral and cellular immunity. They are called $CD4^+$ cells because their main surface marker is CD4. Helper T cells play an intermediate process in the immune response. T cells regulate or "assist" other lymphocytes in their functions [51]. After recognition of antigenic MHC molecules and appropriate co-stimulation, naïve CD4 T cells exit quiescence and undergo clonal expansion. Following cytokine stimulation, different signal transducer and activator of transcription (STAT) family proteins are activated and drive T helper cell differentiation [51].

Thl Cells In 1986, Mosmann et al. divided mouse $CD4^+$ T cells into two functionally distinct and independent subpopulations, Thl and Th2, based on the different cytokine production by $CD4^+$ T cell clones [52]. Later, two subpopulations, Thl and Th2, were also identified in the human $CD4^+$ T cell population. $CD4^+$ Thl cells mainly mediate immune responses related to cytotoxicity and local inflammation and are involved in cellular immunity and the formation of delayed hypersensitivity inflammation [53].

Thl cells are mainly differentiated by naïve T cells under the action of cytokines such as IL-12, which mainly secrete IL-2, IFN-γ, TNF-β, etc. Its main transcription factor is STAT4 in addition to T box factor (T-bet), etc. The surface markers of Thl cells are cytokine receptors IL-12β2R and IL-18R and chemokine receptors CXCR3 and CCR5 [54].

Thl cells play an important role in the resistance to infection by intracellular pathogens (viruses, bacteria, parasites, etc.) [55]. In intracellular bacterial infections, Thl cells preferentially develop and trigger a phagocyte-mediated host defense response. Persistent strong Thl cell responses may be associated with organ-specific autoimmune diseases, contact dermatitis, chronic inflammatory diseases of unknown origin, and delayed hypersensitivity reactions. Their most important executive cells are macrophages [56], in addition to killer $CD8^+$ T cells, IgG-producing B cells, and IFN-γ-secreting $CD4^+$ T cells.

$CD4^+$ Thl cells induce macrophage activation mainly by secreting the macrophage activating factor IFN-γ and by binding to CD40 molecules on the surface of macrophages through surface-expressed CD40L. Activated macrophages show increased expression of surface CD40 molecules and TNF receptors and secretion of TNF-α, thereby increasing macrophage anti-microbial action in concert with IFN-γ.

Thl cells recruit macrophages to the site of infection and stimulate the production of new monocytes in the bone marrow by secreting IL-3 and granulocyte-macrophage colony-stimulating factor (GM-CSF), dilate blood vessels and alter the expression of vascular endothelial cell adhesion molecules [57] (e.g., ICAM-1) by secreting TNF-α and TNF-β, and secrete MCP-1 to attract aggregated macrophages [57].

The auxiliary role of Thl cells is mainly reflected in that the activation and proliferation of $CD8^+$ T cells require the stimulation of IL-2 secreted by $CD4^+$ T cells, while Th1 mainly secretes IL-2, without which the clonal expansion of $CD8^+$ CTL is not possible, and IL-2 also improves the killing activity of $CD8^+$ CTL [58]. By secreting IFN-γ and IL-2, $CD4^+$ Th1 cells can assist B cells in producing regulatory antibodies, which are generally of the IgG1, IgG3, and IgE subclasses [59, 60]. The antibodies bind to target cells with corresponding antigens to form immune complexes and then promote the destruction and clearance of target antigens by binding to phagocytes and NK cells that express CR or FcγR.

Th2 Cells Th2 cells, like Th1 cells, were first discovered by Mosmann et al. when they applied the Th cell clonal culture technique to study the mouse $CD4^+$ T cell population [52]. Th2 is mainly differentiated from naïve cells driven by IL-4 and secretes IL-4, IL-5, IL-6, IL-10, etc. [61]. Its important transcription factor is STAT-6, in addition to GATA-3, etc. The surface markers of Th2 cells are mainly cytokine receptor IFN-γ Rα (CD119), chemokine receptor CXCR4 (CD183), CCR3 (CD193), CCR4 (CD194), etc.

The main function of Th2 cells is to stimulate B cell proliferation, differentiation, and antibody production, which are associated with humoral immunity, and in response to helminth infections and environmental allergens, the humoral immune response is mainly mediated by the involvement of Th2 cells. Excessive Th2 cell response may play an important role in allergic atopy in genetic susceptibility. The most important executive cells are mast cells, eosinophil, and basophil, as well as IgE-producing B cells and IL-4/IL-5-secreting $CD4^+$ T cells [62].

Th2 cell–mediated immune response is the main mechanism by which the body clears parasitic infections and the occurrence of allergic diseases [63, 64]. Recent studies have revealed that IL-25 and IL-33 secreted by intestinal mucosal epithelial cells can induce several novel intrinsic immune cells to promote the differentiation and activation of Th2 cells and increase the secretion of IL-4, IL-5, and IL-13 to clear parasites from the body. Th2 promotes the secretion of IgE antibodies by secreting IL-4, IL-6, and IL-10 from B cells, which in turn interact with IgE and then bind to the high-affinity IgE receptor FcεRI [65] on mast cells and basophils, triggering type I hypersensitivity reactions, such as allergic rhinitis, asthma, and atopic dermatitis, among other autoimmune diseases. Th1 and Th2 have a reciprocal role in the regulation of the immune response, and together with numerous cytokines, they form a complex network of molecules that are involved in cellular and humoral immunity.

Th9 Cells Th9 cells are a new population of $CD4^+$ T cells that secrete mainly IL-9 and IL-10, require synergistic stimulation by

IL-4 and TGF-β, and exert pro-inflammatory effects in a mouse model of enteritis. Th9 cells are important immune cells involved in immune response and immunomodulation, and the main effector that exerts immunomodulatory effects is IL-9.

Th9 cells are a class of IL-9$^+$IL-10$^+$Foxp3-effector T cells. Th9 cells do not express other transcription factors known to be specific to Th cells, including T-bet, GATA-3, Foxp3, and RORγt, with GATA-3 being present at the beginning of Th9 differentiation. Th2 cells can be converted to Th9 cells by TGF-β [66], but after this process, Th9 cells selectively secrete IL-9 and IL-10 at high levels instead of the other cytokines typical of Th cells, and the original Th2 cytokines, including IL-4, IL-5, and IL-13, are hardly produced. In the presence of TGF-β, IL-4 also converts TGF-β-induced Foxp3$^+$Treg cells into Th9 cells, which are detected as CD3$^+$CD4$^+$IL-9.

The main effector of immunomodulation by Th9 cells is IL-9, which has the potential to enhance the onset of IgE-mediated inflammatory responses [67]. Acting on inflammation and tissue cells, it plays an important role in the fight against parasitic infections and the induction of allergic diseases, especially allergic asthma. It has been suggested that IL-9 enhances the body's ability to resist murine whipworm infection. IL-9 is expressed and produced at the beginning of the immune response against the parasite [68], and its increased levels cause the proliferation of mast cells in the intestine, increase the levels of IgE and IgG, stimulate the contraction of intestinal muscles, and promote the elimination of whipworms from the intestine.

The specific mechanism of action of IL-9 is not well understood and needs to be further explored. In allergic diseases, especially allergic asthma, a large amount of IL-9 secreted by Th9 cells acts on different inflammatory cells, such as lymphocytes, eosinophils, mast cells, and neutrophils, causing an increase in the number of inflammatory cells and secreting some inflammatory mediators to promote the development of inflammation [69]. In addition to acting on inflammatory cells, IL-9 also acts on tissue cells such as smooth muscle cells and airway epithelial cells, which secrete pro-inflammatory factors and appear in large numbers as cup cells, secreting large amounts of mucus and playing a role in the development of airway hyperresponsiveness. Tumor-specific CD4 Th9 cells can eradicate large established mouse tumors. The antitumor efficacy of Th9 depends on IL-9 and the upregulation of Eomes and Traf6 [70].

Th17 Cells Th17 cells also differentiate from common naïve CD4$^+$ T cells, in which cytokine species and the balance between cytokines have an important regulatory role in Th cell differentiation [71]. Naïve CD4$^+$ T cells begin to differentiate toward the

Th17 subpopulation stimulated by the TCR pathway and cytokines (mainly IL-6, TGF-β), followed by IL-21 (mainly produced by paracrine and autocrine pathways) to promote the proliferation of the Th17 cell subpopulation and IL-23 to maintain the activity of the Th17 subpopulation and induce its secretion of IL-17, resulting in a pro-inflammatory response.

When the body is in a steady state or without inflammatory injury, the immune system produces TGF-β to inhibit the proliferation of effector T cells and induce Treg cell expression, thus maintaining the body's immune tolerance [72]. However, in the presence of infection or inflammation, the acute phase protein IL-6 is produced in large quantities to inhibit Treg cell expression and, together with TGF-β, induce the differentiation of Th17 cells. IL-23 promotes the differentiation of Th17 cells, while IFN-γ and IL4 inhibit the differentiation of Th17 cells.

Th17 cells are characterized by intracellular expression of RORγt (steroid receptor–type nuclear receptor [73], which is a Th17 cell–specific transcriptional regulator), in addition to a variety of cytokines, including IL-17A, IL-17F, IL-21, and IL-22. Th17 cells have been lacking a clear surface marker, and the current approach, commonly adopted in studies, was proposed in 2009, and that secretion of IL-17A is a phenotypic characteristic of human Th17 cells.

Th17 cells are an inflammatory cell lineage distinct from Th1 and Th2. Th17 are involved in both inflammatory responses, infectious diseases, and autoimmune diseases and play an important role in intrinsic immunity by stimulating epithelial cells to secrete antimicrobial substances such as defensins, as well as recruiting and activating neutrophils. Since Th17 cells produce large amounts of IL-17A, the majority of Th17 cell–mediated effects are mediated by this cytokine. In addition to IL-17A, Th17 cells also produce IL-17F. Both cytokines have similar functions, promoting T cell activation and stimulating epithelial cells, endothelial cells, and fibroblasts to produce a variety of cytokines (e.g., TNF, IL-1β), chemokines (e.g., CXCL1, CXCL8, etc.), and metalloproteinases; it is an early stage of T cell–induced inflammatory response. After binding to the receptor, IL-17 exerts its biological effects through the MAP kinase pathway and nuclear transcription factor kB (NF-κB) pathway. In addition to these, IL-17A and IL-17F are key factors in the recruitment, activation, and migration of neutrophils. IL-17A and IL-17F recruit neutrophils to the site of pathogen infection (inflammation) and promote pathogen elimination, which is the most important biological function of Th17.

In addition to these two cytokines, Th17 cells also secrete IL-21 and IL-22, both of which are not specific to Th17 cells, but they tend to be secreted in Th17 cells. IL-21 exerts its biological effects by interacting with other immune cells (e.g., B cells); in addition, it regulates the differentiation of Th17 cells in an autocrine form. IL-22, on the other hand, exerts its biological effects by inducing the production of pro-inflammatory cytokines and chemokines by other cells after binding to its receptor (a complex of IL-10R2 and IL-22R chains).

Th17 cells and iTreg cells share a common precursor cell (naïve CD4 T cells) and require common TGF-β signaling for initial differentiation. However, terminally differentiated cells achieve opposite functions: Th17 cells cause autoimmunity and inflammation, while Treg cells suppress immunity and maintain immune homeostasis [72, 74]. Since the differentiation of Th17 and Treg cells is mutually regulated by shared and distinct cytokines, and since each subpopulation can be converted to the other under certain inflammatory conditions, the Th17/Treg balance plays a crucial role in many autoimmune diseases. Patients with rheumatoid arthritis (RA), psoriasis, multiple sclerosis (MS), and inflammatory bowel disease (IBD) have elevated Th17/Treg ratios [75, 76]. RA is characterized by chronic inflammation of the synovial membrane, leading to the destruction of bone and cartilage. Th17 cells are thought to be a major driver of pathogenesis, triggering inflammation by activating multiple cell types within the inflamed joint, including macrophages, DCs, and synovial cells) in inflamed joints. Many therapeutic approaches aimed at correcting the balance have been shown to be effective, such as those using monoclonal antibodies to neutralize Th17-related cytokines, including IL-6, TNF-α, IL-17, and IL-23, that have been quite successful [77]. Monoclonal antibodies against human IL-6R, tolimumab, and sarilumab affect the Th17/Treg cell ratio by decreasing Th17 but increasing Treg cell levels in RA patients [78, 79]. Digoxin, an inhibitor of RORγt, ameliorated experimental autoimmune encephalomyelitis (EAE) and collagen-induced arthritis (an animal disease model of RA) by reducing the Th17/Treg ratio [80].

Th22 Cells Th22 is a specific isoform derived from Th17 but turns off many Th17-specific expression patterns [81]. Th22 expresses CCR6, CCR4, and CCR10 and produces IL-22 and IL-13 but not IFN-γ, IL-4, and IL-17. IL-6 and TNF are able to induce differentiation of naïve $CD4^+$ T cells to Th22 cells, and TGF-β inhibits differentiation of Th22. The common markers of Th22 cells include CCR10, IL-22, and CCR6. IL-22 is essential for rapid immune responses against invading pathogens such as bacteria and yeast. Compelling evidence suggests that IL-22 plays an important role in the protection and regeneration of epithelial and keratin-forming cells in the skin, thereby promoting immunity to pathogen entry. In addition, IL-22 induces the expression of pro-inflammatory mediators, including TNF-α, IL-1β, IL-6, IL-8, IL-1, GM-CSF, and G-CSF [82–84]. It is also involved in other physiological procedures including stimulation of antimicrobial peptide production, increased tissue regeneration, and expression of remodeling genes [29].

Th22 cells and IL-22 improve damaged tissue repair and wound healing by enhancing the proliferation and differentiation of epithelial cells and keratin-forming cells. This helps to maintain the integrity of the skin as a broad defense barrier [85]. In addition, IL-22 enhances mucosal immunity by inducing antimicrobial peptides, including human β-defensin 2 and β-defensin 3 [86]. Th22 cells play a regulatory role in the development and progression of many autoimmune diseases. For example, psoriasis, RA, and MS. Th22 and IL-22 are increased in patients with these diseases [87].

Tfh Cells Follicular helper cells (Tfh) are a subset of T cells that assist B cells in antibody production in humoral immunity. Activated by antigens in the T cell zone of lymphoid tissue, they then migrate to the outer edge of the B cell follicle where they encounter antigen-specific B cells [88, 89].

This type of antibody response occurs in the germinal center (GC) of the lymphoid follicle and requires interaction between antigen-specific B and Th cells. Homing of Tfh cells to the GC depends on a two-step chemotaxis process that begins with the downregulation of CCR7 (a receptor for chemokines CCL19 and CCL21, produced in the T cell region of lymphoid tissue) followed by induction of CXCR5, which causes Tfh cells to migrate down a chemokine CXCL13 gradient to the GC [89].

B cell lymphoma 6 protein (Bcl6) is the main transcription factor regulating Tfh cell differentiation [90]. Bcl6 is a transcriptional repressor, and downregulation of CCR7 and upregulation of CXCR5 are driven by Bcl6. The transcription factor Ascl2 controls CXCR5 upregulation independently of Bcl6 and may play a role in the early induction of Tfh cell fate and migration to GC, and the stable induction of Bcl6 is also heavily dependent on the physical interaction of T and B cells. The transcription factor Bhlhe40 is a key cell-intrinsic negative regulator of GC response. In activated CD4 T cells, Bhlhe40 inhibits proliferation and limits the number of TFH cells [91].

Tfh cell differentiation is mediated by IL-21 and IL-6, whereas Tfh cells require STAT3 and C-MAF to be produced. With the activation of Bcl6, Tfh cells begin to secrete IL-21 (Tfh is the predominant IL-21 secreting cell), which proves to be necessary for B cell survival and to promote the GC response. IL-4 and CD40L are other key signals that Tfh cells provide for B cell development. Tfh cells can produce large amounts of dopamine, which is important for a proper GC response. Signal transducer and activator of transcription (STAT) family signaling pathways are essential for Tfh formation. However, due to the multifactorial and multistage nature of Tfh differentiation, each STAT member plays a role in Tfh differentiation but is not fully dependent. With the progressive identification of different Tfh subsets (Tfh1, Tfh2, Tfh17), the precursor Tfh (pre-Tfh) cells share a common developmental program with Th1, Th2, or Th17 cells, and the direction of polarization depends on the dominant cytokine milieu [92].

Due to the flexibility of helper T cells (Th), Tfh cells express the major regulator Bcl-6 and can express major regulators of other lineages, such as T-bet, GATA-binding protein 3 (GATA3), or retinoic acid receptor–associated orphan nuclear receptor γ (RORγT) [93, 94]. These cells expressing dual transcriptional regulators are defined as subpopulations of Tfh cells named Tfh1, Tfh2, and Tfh17. Recently, T follicle–regulated (Tfr) cells have been identified as a new subpopulation of T regulatory (Treg) cells that co-express Bcl-6 and Foxp3. Tfr cells are able to enter B cell follicles and suppress the TfhB cell response in GC [95], which is essential for Tfh differentiation. Tfr differentiation requires CD28 and ICOS expression, whereas PD-1 and CTLA -4 expression appears to inhibit Tfr differentiation [95], although CTLA-4 expression is also an important pathway for Tfr inhibition of B cell responses. Tfr can indirectly inhibit B cell responses (e.g., by regulating Tfh cell function and by regulating B cell metabolism), and Tfr-mediated inhibition can be overcome by IL-21.

Many autoimmune diseases are characterized by the presence of self-reactive autoantibodies [96], and it is thought that GC may drive the pathogenic production of autoantibodies [97]. For example, in an animal model of systemic lupus erythematosus (SLE), spontaneous GC production was associated with increased serum autoantibody concentrations [98]. Furthermore, in animal models of SLE and RA, abnormalities in Tfh cell function can promote disease progression characterized by GC expansion and increased expansion of IL-21 and Tfh cells at extracapsular sites [99, 100].

7 Treg Cells and Immune Homeostasis

Regulatory T cells (Treg) are a subset of T cells with significant immunosuppressive effects. Currently, the recognized surface marker to distinguish Treg cells is $CD4^{+}CD25^{+}FoxP3^{+}$ [101]. It is able to suppress the immune response of other cells and is the primary controller of self-tolerance [102]. Often, its absence or abnormal function leads to the development of autoimmune diseases.

Treg can be broadly classified into two categories depending on the cellular origin, natural regulatory T cells (nTreg), and adaptive regulatory T cells (iTreg) [103]. nTreg are derived from the thymus and are naturally immunosuppressive. nTreg dysregulation leads to autoimmune diseases. iTerg are induced by small doses of antigens or immunosuppressive cytokines from peripheral naïve T cells, including type 1 regulatory T (Tr1) and Th3 cells, mainly secrete IL-10 and TGF-β to play a negative immune regulatory role. Tr1 cells mainly secrete high levels of IL-10 and moderate levels of TGF-β and IFN-γ, but not IL-4 or IL-2. Tr1 cells can suppress the proliferative response of naïve and memory T cells by secreting IL-10 and TGF-β [104, 105]. Th3 cells secrete large amounts of TGF-β, which has a suppressive effect on both Th1 and Th2, and Th3 cells can secrete TGF-β upon antigen [106]. Most autoimmune diseases exhibit defects in the number and function of Treg cells measured in the peripheral blood, such as multiple sclerosis (MS) [107], type 1 diabetes [108], systemic lupus erythematosus (SLE) disease [109], rheumatoid arthritis (RA) [110], and myasthenia gravis [111].

Treg cells play a crucial role in preventing autoimmunity and maintaining self-tolerance. Homeostatic balance is achieved through a combination of soluble cytokine signaling and direct interactions between different cell types. Excessive Treg activity leads to immunodeficiency, chronic infections and cancer, but too little Treg activity leads to autoimmunity and immunopathology and impairs the quality of pathogen-specific responses [112].

Treg cells target a broad spectrum of innate and adaptive immune cells and utilize multiple immunomodulatory mechanisms to limit inflammatory immune responses. Treg cells can enhance suppression by targeting specific effector pathways and/or responses by a variety of mechanisms, including contact-dependent suppression and soluble-mediated suppression [113, 114] (Fig. 2).

Intercellular contact inhibition is one of the main mechanisms by which Treg cells exert immunosuppressive effects; Treg cells express high levels of co-inhibitory receptors, such as CTLA-4, which exert immunosuppressive functions by interacting with ligands on their target cells; CTLA-4 competes with CD28 to bind CD80 and CD86 on APCs, preventing Tconv cell activation and stimulating the expression of indoleamine 2,3-dioxygenase (IDO) in human and murine DC subpopulations [115, 116]. The expression of IDO in DC subpopulations induces the catabolism of the essential amino acid tryptophan to starve T cells. In mice, CTLA-4-specific deletion of Treg cells leads to fatal autoimmune lymphoid proliferative disease [117]. Humans with CTLA-4 mutations also exhibit autosomal dominant autoimmune lymphoid proliferative disease.

Treg also removes antigens bound to MHC II molecules on the surface of DC cells, thereby limiting their present function [118]. LAG-3 on cells interacts with MHC II on immature DCs, leading to the inhibition of DC maturation [119].

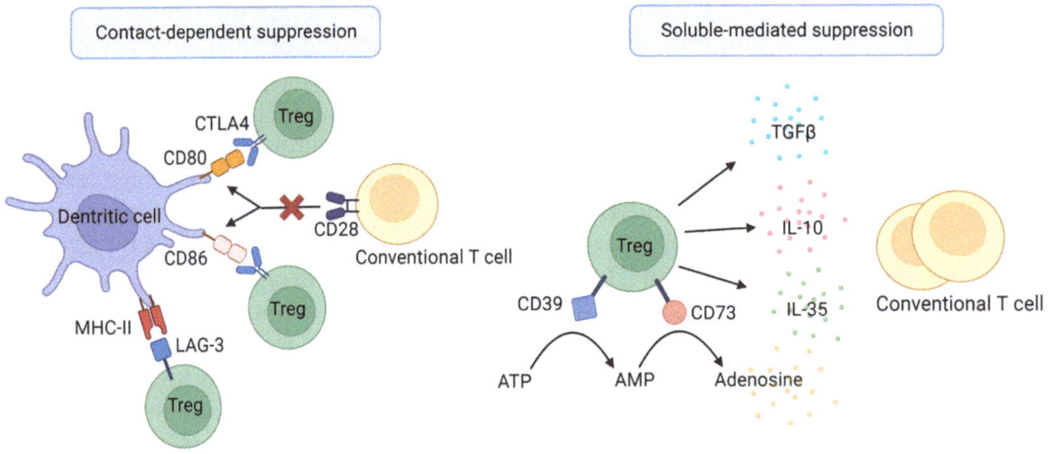

Fig. 2 Contact-dependent suppression and soluble-mediated suppression of Treg. CTLA-4 competes with CD28 for binding to the dendritic cell–expressed costimulatory molecules CD80 and CD86. Compared to CD28, CTLA-4 binds CD80/CD86 with higher affinity, effectively eliminating the costimulatory effects of conventional T cells. lAG-3 is associated with the T cell receptor on the T cell surface and binds to MHC II molecules. This interaction between LAG-3 and MHC II resulted in the inhibition of TCR/CD3-mediated T cell activation. The soluble-mediated suppression of Tregs is the secretion of cytokines such as interleukins IL-10, IL-35, and TGFβ. Tregs can also regulate immunosuppression by producing adenosine, extracellular enzymes CD39 and CD73 on Tregs, or cells closely linked to Tregs that catalyze the breakdown of ATP to adenine

Other mechanisms by which Treg cells may control the immune response through cellular contact include interaction of LAG-3 on Treg cells with MHC II on immature DCs, leading to the inhibition of DC maturation [119]. Antigen-specific Tregs can also physically remove the homologous peptide MHC II complex from the in vivo surface of DCs in a CTLA-4-independent manner [118].

Treg cells secrete several cytokines and can exert their immunosuppressive function through noncontact and bystander effects. TGF-β and IL-10 can act on a wide range of cells and exert a wide range of suppressive functions. For example, they inhibit conventional T cell proliferation and activation, accelerate depletion, inhibit the antigen-presenting function of APCs, and can also induce iTreg cell generation. Mice carrying IL-10-deficient Treg cells develop spontaneous colitis and exhibit higher immune-mediated reactivity in the airways [120]. In human subjects, mutations in the IL-10/IL10R gene are associated with chronic gastrointestinal inflammatory diseases such as ulcerative colitis and Crohn's disease [121]. IL-34 competes with macrophage colony-stimulating factor (CSF-1) for binding to the same receptor, can promote macrophage tolerance, and enhance Treg's own function by autocrine means [122, 123]. IL-35 co-suppresses mainly through upregulation of the receptors PD-1, LAG-3,

TIM-3, and TIGIT and also has similar functions to IL-10 and TGF-β [124].

At high doses, IL-2 promotes the differentiation and proliferation of effector and memory T cells, while at low doses, IL-2 promotes Treg cells. Thus, IL-2 is essential for maintaining immune homeostasis and immune tolerance [125]. In lymph nodes, self-activating T cells produce IL-2, which enhances local Treg proliferation and suppressive function. The resulting microdomains mutually constrain the inputs required for the injury effector response, including CD28 co-stimulation and IL-2 signaling, constituting a negative feedback circuit. As a result of these local constraints, self-activating T cells undergo a brief clonal expansion followed by rapid death [126].

Tregs can also modulate immunosuppression by producing adenosine [127, 128], an unstable and highly inhibitory molecule. Extracellular adenosine accumulates in the body at sites of ischemia and inflammation. In the extracellular space, adenosine is produced by Tregs through the breakdown of ATP. The extracellular enzymes CD39 and CD73 on Tregs, or cells in close association with Tregs, catalyze the breakdown of ATP to adenine. CD73 is widely expressed in activated T cells and other cell types, whereas CD39 expression is largely restricted to Tregs. Sites of inflammation elevated levels of adenosine suppress the immune response by interacting with adenosine A2A receptors on effector T cells or adenosine A2B receptors on antigen-presenting cells [127]. Interaction of extracellular adenosine with either receptor results in elevated intracellular cAMP levels and limits the release of inflammatory cytokines from effector T cells and APCs [127]. Thus, Tregs can catabolize ATP into adenosine via extracellular enzymes, thereby limiting the production of local inflammatory cell motility.

8 Conclusion

T cells coordinate multiple aspects of adaptive immunity, including responses to pathogens, allergens, and tumors. Different T cell subtypes have different functions and work together to maintain immune homeostasis. The immune system appears to maintain immune homeostasis by balancing pathogenic and protective cells. Understanding the mechanisms of tissue specialization of T cell responses will also reveal how long-term health and immune homeostasis are maintained in the face of a variety of antigens, persistent and newly encountered pathogens.

Acknowledgments

This study was supported by the National Natural Science Foundation of China (82022019, 82101414).

References

1. Kumar BV, Connors TJ, Farber DL (2018) Human T cell development, localization, and function throughout life. Immunity 48:202–213
2. Guerder S, Flavell RA (1995) T-cell activation. Two for T. Curr Biol 5:866–868
3. Hedrick SM, Sharp LL (1998) T-cell fate. Immunol Rev 165:95–110
4. Santamaria J, Darrigues J, van Meerwijk JPM, Romagnoli P (2018) Antigen-presenting cells and T-lymphocytes homing to the thymus shape T cell development. Immunol Lett 204:9–15
5. Koch U, Radtke F (2011) Mechanisms of T cell development and transformation. Annu Rev Cell Dev Biol 27:539–562
6. Altin JG, Sloan EK (1997) The role of CD45 and CD45-associated molecules in T cell activation. Immunol Cell Biol 75:430–445
7. Jung Y, Wen L, Altman A, Ley K (2021) CD45 pre-exclusion from the tips of T cell microvilli prior to antigen recognition. Nat Commun 12:3872
8. Malissen B, Ardouin L, Lin SY, Gillet A, Malissen M (1999) Function of the CD3 subunits of the pre-TCR and TCR complexes during T cell development. Adv Immunol 72:103–148
9. Crimeen-Irwin B, Scalzo K, Gloster S, Mottram PL, Plebanski M (2005) Failure of immune homeostasis – the consequences of under and over reactivity. Curr Drug Targets Immune Endocr Metabol Disord 5:413–422
10. Morath A, Schamel WW (2020) αβ and γδ T cell receptors: similar but different. J Leukoc Biol 107:1045–1055
11. Nielsen MM, Witherden DA, Havran WL (2017) γδ T cells in homeostasis and host defence of epithelial barrier tissues. Nat Rev Immunol 17:733–745
12. Rigau M, Ostrouska S, Fulford TS, Johnson DN, Woods K, Ruan Z, McWilliam HEG, Hudson C, Tutuka C, Wheatley AK, Kent SJ, Villadangos JA, Pal B, Kurts C, Simmonds J, Pelzing M, Nash AD, Hammet A, Verhagen AM, Vairo G, Maraskovsky E, Panousis C, Gherardin NA, Cebon J, Godfrey DI, Behren A, Uldrich AP (2020) Butyrophilin 2A1 is essential for phosphoantigen reactivity by γδ T cells. Science 367:eaay5516
13. Dimova T, Brouwer M, Gosselin F, Tassignon J, Leo O, Donner C, Marchant A, Vermijlen D (2015) Effector Vγ9Vδ2 T cells dominate the human fetal γδ T-cell repertoire. Proc Natl Acad Sci U S A 112:E556–E565
14. McVay LD, Carding SR, Bottomly K, Hayday AC (1991) Regulated expression and structure of T cell receptor gamma/delta transcripts in human thymic ontogeny. EMBO J 10:83–91
15. Tan L, Sandrock I, Odak I, Aizenbud Y, Wilharm A, Barros-Martins J, Tabib Y, Borchers A, Amado T, Gangoda L, Herold MJ, Schmidt-Supprian M, Kisielow J, Silva-Santos B, Koenecke C, Hovav A-H, Krebs C, Prinz I, Ravens S (2019) Single-cell transcriptomics identifies the adaptation of Scart1+ Vγ6+ T cells to skin residency as activated effector cells. Cell Rep 27:3657–3671.e4
16. Marchitto MC, Dillen CA, Liu H, Miller RJ, Archer NK, Ortines RV, Alphonse MP, Marusina AI, Merleev AA, Wang Y, Pinsker BL, Byrd AS, Brown ID, Ravipati A, Zhang E, Cai SS, Limjunyawong N, Dong X, Yeaman MR, Simon SI, Shen W, Durum SK, O'Brien RL, Maverakis E, Miller LS (2019) Clonal Vγ6+Vδ4+ T cells promote IL-17-mediated immunity against Staphylococcus aureus skin infection. Proc Natl Acad Sci U S A 116:10917–10926
17. Muñoz-Ruiz M, Sumaria N, Pennington DJ, Silva-Santos B (2017) Thymic determinants of γδ T cell differentiation. Trends Immunol 38:336–344
18. Isailovic N, Daigo K, Mantovani A, Selmi C (2015) Interleukin-17 and innate immunity in infections and chronic inflammation. J Autoimmun 60:1–11
19. Hirota K, Turner J-E, Villa M, Duarte JH, Demengeot J, Steinmetz OM, Stockinger B (2013) Plasticity of Th17 cells in Peyer's patches is responsible for the induction of T cell-dependent IgA responses. Nat Immunol 14:372–379
20. Papotto PH, Reinhardt A, Prinz I, Silva-Santos B (2018) Innately versatile: γδ17 T

cells in inflammatory and autoimmune diseases. J Autoimmun 87:26–37

21. Schirmer L, Rothhammer V, Hemmer B, Korn T (2013) Enriched CD161high CCR6+ γδ T cells in the cerebrospinal fluid of patients with multiple sclerosis. JAMA Neurol 70:345–351
22. Kurasawa K, Hirose K, Sano H, Endo H, Shinkai H, Nawata Y, Takabayashi K, Iwamoto I (2000) Increased interleukin-17 production in patients with systemic sclerosis. Arthritis Rheum 43:2455–2463
23. Prager I, Liesche C, van Ooijen H, Urlaub D, Verron Q, Sandström N, Fasbender F, Claus M, Eils R, Beaudouin J, Önfelt B, Watzl C (2019) NK cells switch from granzyme B to death receptor-mediated cytotoxicity during serial killing. J Exp Med 216: 2113–2127
24. Voskoboinik I, Whisstock JC, Trapani JA (2015) Perforin and granzymes: function, dysfunction and human pathology. Nat Rev Immunol 15:388–400
25. Etxeberria I, Glez-Vaz J, Teijeira Á, Melero I (2020) New emerging targets in cancer immunotherapy: CD137/4-1BB costimulatory axis. ESMO Open 4:e000733
26. den Braber I, Mugwagwa T, Vrisekoop N, Westera L, Mögling R, de Boer AB, Willems N, Schrijver EHR, Spierenburg G, Gaiser K, Mul E, Otto SA, Ruiter AFC, Ackermans MT, Miedema F, Borghans JAM, de Boer RJ, Tesselaar K (2012) Maintenance of peripheral naive T cells is sustained by thymus output in mice but not humans. Immunity 36:288–297
27. Cieri N, Camisa B, Cocchiarella F, Forcato M, Oliveira G, Provasi E, Bondanza A, Bordignon C, Peccatori J, Ciceri F, Lupo-Stanghellini MT, Mavilio F, Mondino A, Bicciato S, Recchia A, Bonini C (2013) IL-7 and IL-15 instruct the generation of human memory stem T cells from naive precursors. Blood 121:573–584
28. Cimbro R, Vassena L, Arthos J, Cicala C, Kehrl JH, Park C, Sereti I, Lederman MM, Fauci AS, Lusso P (2012) IL-7 induces expression and activation of integrin α4β7 promoting naive T-cell homing to the intestinal mucosa. Blood 120:2610–2619
29. Surh CD, Sprent J (2008) Homeostasis of naive and memory T cells. Immunity 29: 848–862
30. Vivien L, Benoist C, Mathis D (2001) T lymphocytes need IL-7 but not IL-4 or IL-6 to survive in vivo. Int Immunol 13:763–768
31. Tan JT, Dudl E, LeRoy E, Murray R, Sprent J, Weinberg KI, Surh CD (2001) IL-7 is critical for homeostatic proliferation and survival of naive T cells. Proc Natl Acad Sci U S A 98: 8732–8737
32. Raeber ME, Zurbuchen Y, Impellizzieri D, Boyman O (2018) The role of cytokines in T-cell memory in health and disease. Immunol Rev 283:176–193
33. Takamura S (2020) Divergence of tissue-memory T cells: distribution and function-based classification. Cold Spring Harb Perspect Biol 12:a037762
34. Mueller SN, Gebhardt T, Carbone FR, Heath WR (2013) Memory T cell subsets, migration patterns, and tissue residence. Annu Rev Immunol 31:137–161
35. Turner DL, Bickham KL, Thome JJ, Kim CY, D'Ovidio F, Wherry EJ, Farber DL (2014) Lung niches for the generation and maintenance of tissue-resident memory T cells. Mucosal Immunol 7:501–510
36. Teijaro JR, Turner D, Pham Q, Wherry EJ, Lefrançois L, Farber DL (2011) Cutting edge: tissue-retentive lung memory CD4 T cells mediate optimal protection to respiratory virus infection. J Immunol 187: 5510–5514
37. Herndler-Brandstetter D, Landgraf K, Jenewein B, Tzankov A, Brunauer R, Brunner S, Parson W, Kloss F, Gassner R, Lepperdinger G, Grubeck-Loebenstein B (2011) Human bone marrow hosts polyfunctional memory CD4+ and CD8+ T cells with close contact to IL-15-producing cells. J Immunol 186:6965–6971
38. Radbruch A, McGrath MA, Siracusa F, Hoffmann U, Sercan-Alp Ö, Hutloff A, Tokoyoda K, Chang H-D, Dong J (2021) Homeostasis and durability of T-cell memory-the resting and the restless T-cell memory. Cold Spring Harb Perspect Biol 13: a038083
39. Becker TC, Coley SM, Wherry EJ, Ahmed R (2005) Bone marrow is a preferred site for homeostatic proliferation of memory CD8 T cells. J Immunol 174:1269–1273
40. Kondrack RM, Harbertson J, Tan JT, McBreen ME, Surh CD, Bradley LM (2003) Interleukin 7 regulates the survival and generation of memory CD4 cells. J Exp Med 198: 1797–1806
41. Li J, Huston G, Swain SL (2003) IL-7 promotes the transition of CD4 effectors to persistent memory cells. J Exp Med 198:1807–1815

42. Becker TC, Wherry EJ, Boone D, Murali-Krishna K, Antia R, Ma A, Ahmed R (2002) Interleukin 15 is required for proliferative renewal of virus-specific memory CD8 T cells. J Exp Med 195:1541–1548
43. Goldrath AW, Sivakumar PV, Glaccum M, Kennedy MK, Bevan MJ, Benoist C, Mathis D, Butz EA (2002) Cytokine requirements for acute and Basal homeostatic proliferation of naive and memory CD8+ T cells. J Exp Med 195:1515–1522
44. Kolan SS, Li G, Wik JA, Malachin G, Guo S, Kolan P, Skålhegg BS (2020) Cellular metabolism dictates T cell effector function in health and disease. Scand J Immunol 92:e12956
45. Weigelin B, Krause M, Friedl P (2011) Cytotoxic T lymphocyte migration and effector function in the tumor microenvironment. Immunol Lett 138:19–21
46. Mosmann TR, Li L, Sad S (1997) Functions of CD8 T-cell subsets secreting different cytokine patterns. Semin Immunol 9:87–92
47. Cao X, Cai SF, Fehniger TA, Song J, Collins LI, Piwnica-Worms DR, Ley TJ (2007) Granzyme B and perforin are important for regulatory T cell-mediated suppression of tumor clearance. Immunity 27:635–646
48. Jonsson AH, Zhang F, Dunlap G, Gomez-Rivas E, Watts GFM, Faust HJ, Rupani KV, Mears JR, Meednu N, Wang R, Keras G, Coblyn JS, Massarotti EM, Todd DJ, Anolik JH, McDavid A, Wei K, Rao DA, Raychaudhuri S, Brenner MB (2022) Granzyme K+ CD8 T cells form a core population in inflamed human tissue. Sci Transl Med 14: eabo0686
49. Flores-Mendoza G, Rodríguez-Rodríguez N, Rubio RM, Madera-Salcedo IK, Rosetti F, Crispín JC (2021) Fas/FasL signaling regulates CD8 expression during exposure to self-antigens. Front Immunol 12:635862
50. Yajima T, Hoshino K, Muranushi R, Mogi A, Onozato R, Yamaki E, Kosaka T, Tanaka S, Shirabe K, Yoshikai Y, Kuwano H (2019) Fas/FasL signaling is critical for the survival of exhausted antigen-specific CD8+ T cells during tumor immune response. Mol Immunol 107:97–105
51. Saravia J, Chapman NM, Chi H (2019) Helper T cell differentiation. Cell Mol Immunol 16:634–643
52. Zhu J, Paul WE (2008) CD4 T cells: fates, functions, and faults. Blood 112:1557–1569
53. Romagnani S (1999) Th1/Th2 cells. Inflamm Bowel Dis 5:285–294
54. Krueger PD, Goldberg MF, Hong S-W, Osum KC, Langlois RA, Kotov DI, Dileepan T, Jenkins MK (2021) Two sequential activation modules control the differentiation of protective T helper-1 (Th1) cells. Immunity 54:687–701.e4
55. Bagheri N, Salimzadeh L, Shirzad H (2018) The role of T helper 1-cell response in Helicobacter pylori-infection. Microb Pathog 123:1–8
56. Ma J, Chen T, Mandelin J, Ceponis A, Miller NE, Hukkanen M, Ma GF, Konttinen YT (2003) Regulation of macrophage activation. Cell Mol Life Sci 60:2334–2346
57. Ogawa A, Yoshizaki A, Yanaba K, Ogawa F, Hara T, Muroi E, Takenaka M, Shimizu K, Hasegawa M, Fujimoto M, Tedder TF, Sato S (2010) The differential role of L-selectin and ICAM-1 in Th1-type and Th2-type contact hypersensitivity. J Invest Dermatol 130: 1558–1570
58. Snell LM, Osokine I, Yamada DH, De la Fuente JR, Elsaesser HJ, Brooks DG (2016) Overcoming CD4 Th1 cell fate restrictions to sustain antiviral CD8 T cells and control persistent virus infection. Cell Rep 16:3286–3296
59. Zoldan K, Ehrlich S, Killmer S, Wild K, Smits M, Russ M, Globig A-M, Hofmann M, Thimme R, Boettler T (2021) Th1-biased hepatitis C virus-specific follicular T helper-like cells effectively support B cells after antiviral therapy. Front Immunol 12: 742061
60. Smith KM, Brewer JM, Rush CM, Riley J, Garside P (2004) In vivo generated Th1 cells can migrate to B cell follicles to support B cell responses. J Immunol 173:1640–1646
61. Ruterbusch M, Pruner KB, Shehata L, Pepper M (2020) In vivo CD4+ T cell differentiation and function: revisiting the Th1/Th2 paradigm. Annu Rev Immunol 38:705–725
62. Aalberse RC, Platts-Mills TAE (2004) How do we avoid developing allergy: modifications of the TH2 response from a B-cell perspective. J Allergy Clin Immunol 113:983–986
63. Ogulur I, Pat Y, Ardicli O, Barletta E, Cevhertas L, Fernandez-Santamaria R, Huang M, Bel Imam M, Koch J, Ma S, Maurer DJ, Mitamura Y, Peng Y, Radzikowska U, Rinaldi AO, Rodriguez-Coira J, Satitsuksanoa P, Schneider SR, Wallimann A, Zhakparov D, Ziadlou R, Brüggen M-C, van de Veen W, Sokolowska M, Baerenfaller K, Zhang L, Akdis M, Akdis CA (2021) Advances and highlights in biomarkers of allergic diseases. Allergy 76:3659–3686

64. Nakayama T, Hirahara K, Onodera A, Endo Y, Hosokawa H, Shinoda K, Tumes DJ, Okamoto Y (2017) Th2 cells in health and disease. Annu Rev Immunol 35:53–84
65. Turner H, Kinet JP (1999) Signalling through the high-affinity IgE receptor Fc epsilonRI. Nature 402:B24–B30
66. Abdelaziz MH, Wang H, Cheng J, Xu H (2020) Th2 cells as an intermediate for the differentiation of naïve T cells into Th9 cells, associated with the Smad3/Smad4 and IRF4 pathway. Exp Ther Med 19:1947–1954
67. Olson MR, Kaplan MH (2019) TH9 immunodeficiency in patients with hyper-IgE syndrome. J Allergy Clin Immunol 143:935–936
68. Licona-Limón P, Arias-Rojas A, Olguín-Martínez E (2017) IL-9 and Th9 in parasite immunity. Semin Immunopathol 39:29–38
69. Fu Y, Wang J, Zhou B, Pajulas A, Gao H, Ramdas B, Koh B, Ulrich BJ, Yang S, Kapur R, Renauld J-C, Paczesny S, Liu Y, Tighe RM, Licona-Limón P, Flavell RA, Takatsuka S, Kitamura D, Tepper RS, Sun J, Kaplan MH (2022) An IL-9-pulmonary macrophage axis defines the allergic lung inflammatory environment. Sci Immunol 7: eabi9768
70. Lu Y, Wang Q, Xue G, Bi E, Ma X, Wang A, Qian J, Dong C, Yi Q (2018) Th9 cells represent a unique subset of CD4+ T cells endowed with the ability to eradicate advanced tumors. Cancer Cell 33:1048–1060.e7
71. Sallusto F, Zielinski CE, Lanzavecchia A (2012) Human Th17 subsets. Eur J Immunol 42:2215–2220
72. Zhang W, Liu X, Zhu Y, Liu X, Gu Y, Dai X, Li B (2021) Transcriptional and posttranslational regulation of Th17/Treg balance in health and disease. Eur J Immunol 51:2137–2150
73. Kumar R, Theiss AL, Venuprasad K (2021) RORγt protein modifications and IL-17-mediated inflammation. Trends Immunol 42:1037–1050
74. Lee GR (2018) The balance of Th17 versus treg cells in autoimmunity. Int J Mol Sci 19: 730
75. Bettelli E, Korn T, Oukka M, Kuchroo VK (2008) Induction and effector functions of T (H)17 cells. Nature 453:1051–1057
76. Lock C, Hermans G, Pedotti R, Brendolan A, Schadt E, Garren H, Langer-Gould A, Strober S, Cannella B, Allard J, Klonowski P, Austin A, Lad N, Kaminski N, Galli SJ, Oksenberg JR, Raine CS, Heller R, Steinman L (2002) Gene-microarray analysis of multiple sclerosis lesions yields new targets validated in autoimmune encephalomyelitis. Nat Med 8:500–508
77. Noack M, Miossec P (2014) Th17 and regulatory T cell balance in autoimmune and inflammatory diseases. Autoimmun Rev 13: 668–677
78. Kikuchi J, Hashizume M, Kaneko Y, Yoshimoto K, Nishina N, Takeuchi T (2015) Peripheral blood CD4(+)CD25(+)CD127 (low) regulatory T cells are significantly increased by tocilizumab treatment in patients with rheumatoid arthritis: increase in regulatory T cells correlates with clinical response. Arthritis Res Ther 17:10
79. Samson M, Audia S, Janikashvili N, Ciudad M, Trad M, Fraszczak J, Ornetti P, Maillefert J-F, Miossec P, Bonnotte B (2012) Brief report: inhibition of interleukin-6 function corrects Th17/Treg cell imbalance in patients with rheumatoid arthritis. Arthritis Rheum 64:2499–2503
80. Lee J, Baek S, Lee J, Lee J, Lee D-G, Park M-K, Cho M-L, Park S-H, Kwok S-K (2015) Digoxin ameliorates autoimmune arthritis via suppression of Th17 differentiation. Int Immunopharmacol 26:103–111
81. Doulabi H, Masoumi E, Rastin M, Foolady Azarnaminy A, Esmaeili S-A, Mahmoudi M (2022) The role of Th22 cells, from tissue repair to cancer progression. Cytokine 149: 155749
82. Dudakov JA, Hanash AM, van den Brink MRM (2015) Interleukin-22: immunobiology and pathology. Annu Rev Immunol 33: 747–785
83. Andoh A, Zhang Z, Inatomi O, Fujino S, Deguchi Y, Araki Y, Tsujikawa T, Kitoh K, Kim-Mitsuyama S, Takayanagi A, Shimizu N, Fujiyama Y (2005) Interleukin-22, a member of the IL-10 subfamily, induces inflammatory responses in colonic subepithelial myofibroblasts. Gastroenterology 129:969–984
84. Abadi AH, Mahdavi M, Khaledi A, Esmaeili S-A, Esmaeili D, Sahebkar A (2018) Study of serum bactericidal and splenic activity of Total-OMP- CagA combination from Brucella abortus and Helicobacter pylori in BALB/c mouse model. Microb Pathog 121: 100–105
85. Ouyang W, O'Garra A (2019) IL-10 family cytokines IL-10 and IL-22: from basic science to clinical translation. Immunity 50:871–891
86. Aujla SJ, Chan YR, Zheng M, Fei M, Askew DJ, Pociask DA, Reinhart TA, McAllister F, Edeal J, Gaus K, Husain S, Kreindler JL, Dubin PJ, Pilewski JM, Myerburg MM, Mason CA, Iwakura Y, Kolls JK (2008)

IL-22 mediates mucosal host defense against Gram-negative bacterial pneumonia. Nat Med 14:275–281

87. Jiang Q, Yang G, Xiao F, Xie J, Wang S, Lu L, Cui D (2021) Role of Th22 cells in the pathogenesis of autoimmune diseases. Front Immunol 12:688066
88. Olatunde AC, Hale JS, Lamb TJ (2021) Cytokine-skewed Tfh cells: functional consequences for B cell help. Trends Immunol 42: 536–550
89. Crotty S (2011) Follicular helper CD4 T cells (TFH). Annu Rev Immunol 29:621–663
90. Choi J, Crotty S (2021) Bcl6-mediated transcriptional regulation of follicular helper T cells (TFH). Trends Immunol 42:336–349
91. Rauschmeier R, Reinhardt A, Gustafsson C, Glaros V, Artemov AV, Dunst J, Taneja R, Adameyko I, Månsson R, Busslinger M, Kreslavsky T (2022) Bhlhe40 function in activated B and TFH cells restrains the GC reaction and prevents lymphomagenesis. J Exp Med 219:e20211406
92. Ueno H, Banchereau J, Vinuesa CG (2015) Pathophysiology of T follicular helper cells in humans and mice. Nat Immunol 16:142–152
93. Nakayamada S, Takahashi H, Kanno Y, O'Shea JJ (2012) Helper T cell diversity and plasticity. Curr Opin Immunol 24:297–302
94. Ma X, Nakayamada S (2021) Multi-source pathways of T follicular helper cell differentiation. Front Immunol 12:621105
95. Fonseca VR, Ribeiro F, Graca L (2019) T follicular regulatory (Tfr) cells: dissecting the complexity of Tfr-cell compartments. Immunol Rev 288:112–127
96. Ding T, Su R, Wu R, Xue H, Wang Y, Su R, Gao C, Li X, Wang C (2021) Frontiers of autoantibodies in autoimmune disorders: crosstalk between Tfh/Tfr and regulatory B cells. Front Immunol 12:641013
97. Wei X, Niu X (2022) T follicular helper cells in autoimmune diseases. J Autoimmun 134: 102976
98. Zhang Q, Xiang L, Zaman MH, Dong W, He G, Deng G-M (2019) Predominant role of immunoglobulin G in the pathogenesis of splenomegaly in murine lupus. Front Immunol 10:3020
99. Wichner K, Stauss D, Kampfrath B, Krüger K, Müller G, Rehm A, Lipp M, Höpken UE (2016) Dysregulated development of IL-17- and IL-21-expressing follicular helper T cells and increased germinal center formation in the absence of RORγt. FASEB J 30:761–774
100. Lu J, Wu J, Xia X, Peng H, Wang S (2021) Follicular helper T cells: potential therapeutic targets in rheumatoid arthritis. Cell Mol Life Sci 78:5095–5106
101. Ohkura N, Hamaguchi M, Morikawa H, Sugimura K, Tanaka A, Ito Y, Osaki M, Tanaka Y, Yamashita R, Nakano N, Huehn J, Fehling HJ, Sparwasser T, Nakai K, Sakaguchi S (2012) T cell receptor stimulation-induced epigenetic changes and Foxp3 expression are independent and complementary events required for Treg cell development. Immunity 37:785–799
102. Shevyrev D, Tereshchenko V (2019) Treg heterogeneity, function, and homeostasis. Front Immunol 10:3100
103. Benoist C, Mathis D (2012) Treg cells, life history, and diversity. Cold Spring Harb Perspect Biol 4:a007021
104. Song Y, Wang N, Chen L, Fang L (2021) Tr1 cells as a key regulator for maintaining immune homeostasis in transplantation. Front Immunol 12:671579
105. Battaglia M, Gregori S, Bacchetta R, Roncarolo M-G (2006) Tr1 cells: from discovery to their clinical application. Semin Immunol 18: 120–127
106. Carrier Y, Yuan J, Kuchroo VK, Weiner HL (2007) Th3 cells in peripheral tolerance. II. TGF-beta-transgenic Th3 cells rescue IL-2-deficient mice from autoimmunity. J Immunol 178:172–178
107. Viglietta V, Baecher-Allan C, Weiner HL, Hafler DA (2004) Loss of functional suppression by CD4+CD25+ regulatory T cells in patients with multiple sclerosis. J Exp Med 199:971–979
108. Haseda F, Imagawa A, Murase-Mishiba Y, Terasaki J, Hanafusa T (2013) $CD4^+$ $CD45RA^-$ FoxP3high activated regulatory T cells are functionally impaired and related to residual insulin-secreting capacity in patients with type 1 diabetes. Clin Exp Immunol 173:207–216
109. Bonelli M, Savitskaya A, von Dalwigk K, Steiner CW, Aletaha D, Smolen JS, Scheinecker C (2008) Quantitative and qualitative deficiencies of regulatory T cells in patients with systemic lupus erythematosus (SLE). Int Immunol 20:861–868
110. van Roon JAG, Hartgring SAY, van der Wurff-Jacobs KMG, Bijlsma JWJ, Lafeber FPJG (2010) Numbers of CD25+Foxp3+ T cells that lack the IL-7 receptor are increased intra-articularly and have impaired suppressive function in RA patients. Rheumatology (Oxford) 49:2084–2089
111. Thiruppathi M, Rowin J, Li Jiang Q, Sheng JR, Prabhakar BS, Meriggioli MN (2012)

Functional defect in regulatory T cells in myasthenia gravis. Ann N Y Acad Sci 1274:68–76

112. Smigiel KS, Srivastava S, Stolley JM, Campbell DJ (2014) Regulatory T-cell homeostasis: steady-state maintenance and modulation during inflammation. Immunol Rev 259:40–59
113. Arpaia N, Green JA, Moltedo B, Arvey A, Hemmers S, Yuan S, Treuting PM, Rudensky AY (2015) A distinct function of regulatory T cells in tissue protection. Cell 162:1078–1089
114. Kanamori M, Nakatsukasa H, Okada M, Lu Q, Yoshimura A (2016) Induced regulatory T cells: their development, stability, and applications. Trends Immunol 37:803–811
115. Qureshi OS, Zheng Y, Nakamura K, Attridge K, Manzotti C, Schmidt EM, Baker J, Jeffery LE, Kaur S, Briggs Z, Hou TZ, Futter CE, Anderson G, Walker LSK, Sansom DM (2011) Trans-endocytosis of CD80 and CD86: a molecular basis for the cell-extrinsic function of CTLA-4. Science 332:600–603
116. Onishi Y, Fehervari Z, Yamaguchi T, Sakaguchi S (2008) Foxp3+ natural regulatory T cells preferentially form aggregates on dendritic cells in vitro and actively inhibit their maturation. Proc Natl Acad Sci U S A 105:10113–10118
117. Yan Z, Garg SK, Banerjee R (2010) Regulatory T cells interfere with glutathione metabolism in dendritic cells and T cells. J Biol Chem 285:41525–41532
118. Akkaya B, Oya Y, Akkaya M, Al Souz J, Holstein AH, Kamenyeva O, Kabat J, Matsumura R, Dorward DW, Glass DD, Shevach EM (2019) Regulatory T cells mediate specific suppression by depleting peptide-MHC class II from dendritic cells. Nat Immunol 20:218–231
119. Liang B, Workman C, Lee J, Chew C, Dale BM, Colonna L, Flores M, Li N, Schweighoffer E, Greenberg S, Tybulewicz V, Vignali D, Clynes R (2008) Regulatory T cells inhibit dendritic cells by lymphocyte activation gene-3 engagement of MHC class II. J Immunol 180:5916–5926
120. Rubtsov YP, Rasmussen JP, Chi EY, Fontenot J, Castelli L, Ye X, Treuting P, Siewe L, Roers A, Henderson WR, Muller W, Rudensky AY (2008) Regulatory T cell-derived interleukin-10 limits inflammation at environmental interfaces. Immunity 28:546–558
121. Chaudhry A, Samstein RM, Treuting P, Liang Y, Pils MC, Heinrich J-M, Jack RS, Wunderlich FT, Brüning JC, Müller W, Rudensky AY (2011) Interleukin-10 signaling in regulatory T cells is required for suppression of Th17 cell-mediated inflammation. Immunity 34:566–578
122. Freuchet A, Salama A, Bézie S, Tesson L, Rémy S, Humeau R, Règue H, Sérazin C, Flippe L, Peterson P, Vimond N, Usal C, Ménoret S, Heslan J-M, Duteille F, Blanchard F, Giral M, Colonna M, Anegon I, Guillonneau C (2022) IL-34 deficiency impairs FOXP3+ Treg function in a model of autoimmune colitis and decreases immune tolerance homeostasis. Clin Transl Med 12:e988
123. Guillonneau C, Bézie S, Anegon I (2017) Immunoregulatory properties of the cytokine IL-34. Cell Mol Life Sci 74:2569–2586
124. Collison LW, Workman CJ, Kuo TT, Boyd K, Wang Y, Vignali KM, Cross R, Sehy D, Blumberg RS, Vignali DAA (2007) The inhibitory cytokine IL-35 contributes to regulatory T-cell function. Nature 450:566–569
125. Spolski R, Li P, Leonard WJ (2018) Biology and regulation of IL-2: from molecular mechanisms to human therapy. Nat Rev Immunol 18:648–659
126. Wong HS, Park K, Gola A, Baptista AP, Miller CH, Deep D, Lou M, Boyd LF, Rudensky AY, Savage PA, Altan-Bonnet G, Tsang JS, Germain RN (2021) A local regulatory T cell feedback circuit maintains immune homeostasis by pruning self-activated T cells. Cell 184:3981–3997.e22
127. Sauer AV, Brigida I, Carriglio N, Hernandez RJ, Scaramuzza S, Clavenna D, Sanvito F, Poliani PL, Gagliani N, Carlucci F, Tabucchi A, Roncarolo MG, Traggiai E, Villa A, Aiuti A (2012) Alterations in the adenosine metabolism and CD39/CD73 adenosinergic machinery cause loss of Treg cell function and autoimmunity in ADA-deficient SCID. Blood 119:1428–1439
128. Regateiro FS, Cobbold SP, Waldmann H (2013) CD73 and adenosine generation in the creation of regulatory microenvironments. Clin Exp Immunol 171:1–7

Chapter 4

Posttranscriptional Events Orchestrate Immune Homeostasis of CD8^{+} T Cells

Julian J. Freen-van Heeren

Abstract

Maintaining immune homeostasis is instrumental for host health. Immune cells, such as T cells, are instrumental for the eradication of pathogenic bacteria, fungi and viruses. Furthermore, T cells also play a major role in the fight against cancer. Through the formation of immunological memory, a pool of antigen-experienced T cells remains in the body to rapidly protect the host upon reinfection or retransformation. In order to perform their protective function, T cells produce cytolytic molecules, such as granzymes and perforin, and cytokines such as interferon γ and tumor necrosis factor α. Recently, it has become evident that posttranscriptional regulatory events dictate the kinetics and magnitude of cytokine production by murine and human CD8^{+} T cells. Here, the recent literature regarding the role posttranscriptional regulation plays in maintaining immune homeostasis of antigen-experienced CD8^{+} T cells is reviewed.

Key words T cells, Posttranscriptional regulation, Immune homeostasis, Effector function

1 Introduction

Maintaining immune homeostasis is instrumental for host health. Immune cells are key players in the defense against pathogenic bacteria, fungi and viruses. The immune response can be divided into innate and adaptive immune responses. Where innate immune responses are of paramount importance for short-term and immediate protection, long-lasting immunological memory is provided by cells from the adaptive immune system. As part of the adaptive immune system, memory T cells specific for previously encountered pathogens remain in the body, ready to be rapidly reactivated upon pathogen reencounter. Compared to naïve T cells, memory T cells respond more rapidly to infection and can exert their effector function more efficiently. As a result, reencountering the same

Vijay Kumar (ed.), *Immune Homeostasis: Methods and Protocols*, Methods in Molecular Biology, vol. 2782,
https://doi.org/10.1007/978-1-0716-3754-8_4,

invading pathogen results in more rapid clearance compared to primary infections. Furthermore, T cells also play a major role in the fight against cancer.

T cells originate in the thymus (Fig. 1). $CD34^+$ bone marrow–borne precursor cells populate the thymus after trafficking to the organ via the blood vessels [1]. These cells are pluripotent and can differentiate into cells of both the myeloid and lymphoid lineage [2]. Once in the thymus, commitment to the lymphocyte lineage occurs upon receiving the correct signals to do so. The transcription factor NOTCH1 is important for inducing T cell commitment in lymphocyte precursors [3]. At this stage, T cell precursors are $CD4^-CD8^-$ and express CD1a [2]. Signals from homeostatic cytokines, such as IL-7, ensure the survival of progenitor T cells [4]. Next, cells that will be conventional T cells acquire CD4 expression and commit to the αβ T cell lineage [5]. The selection of intact β T cell receptor (TCR) chains ensures that defective T cells are effectively removed [2]. During the process of αβ commitment, T cell precursors also acquire the expression of CD8, resulting in $CD4^+CD8^+$ cells. Recombination-activating gene, or RAG,-mediated TCR α-chain rearrangement subsequently results in the

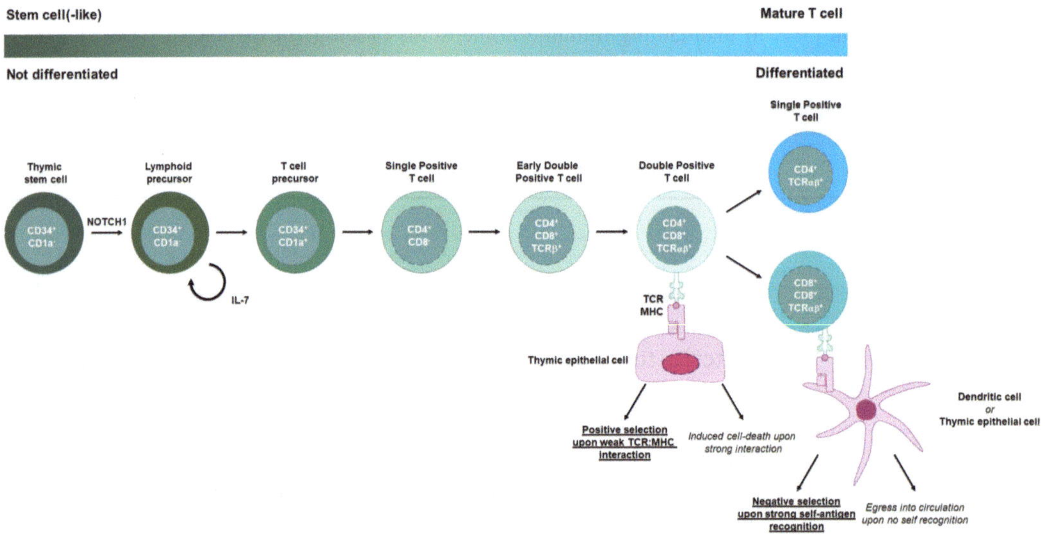

Fig. 1 T cell development in the thymus. Upon committing to the lymphoid line by, for example, NOTCH1 activation, thymic stem cells become lymphoid precursors. These precursor cells maintain themselves through homeostatic signaling, i.e. through sensing of IL-7. Through a sequential maturation process, early T cells first acquire CD4 expression (the single positive stage) and then simultaneously acquire CD8 and TCRβ expression. Next, TCR rearrangement occurs, resulting in the expression of a functional TCR. Through a positive selection process, only T cells that are capable of weakly recognizing major histocompatibility complex, or MHC, expressed on thymic epithelial cells survive. These positively selected cells then become either helper $CD4^+$ T cells or cytotoxic $CD8^+$ T cells, which are negatively selected to not bind self-antigens before allowing to egress into the circulation as fully matured T cells

production of TCRs. These TCRs are first tested to be able to recognize presented antigens through a weak interaction in a process termed positive selection. Next, to maintain immune homeostasis and prevent unwanted immune activation against self-antigens, T cells with a TCR that recognize self-antigens are removed through negative selection [6, 7]. After specialization into either CD4 or CD8 single positive T cells and further maturation, naïve T cells exit the thymus and circulate in the peripheral blood and the lymphatic system, patrolling the body in search of their cognate antigen. Of note, it has recently been shown that mouse T cells have the capacity to surpass the lifespan of the organism by almost four times [8, 9], showing the importance of tightly regulating T cell function to maintain host health.

For naïve T cells to combat pathogens, they first need to be activated. Naïve T cells require three signals to be able to exert their effector function. First, T cells require the recognition of the cognate antigen their TCR is specific for [10]. Second, co-stimulatory signals, for instance, through the triggering of costimulatory receptors such as cluster of differentiation (CD)28 or CD27, are important to enhance TCR-mediated activation [11–15]. Lastly, T cells require cytokine and/or chemokine signals for optimal activation and differentiation [16, 17]. Together, these three signals promote T cell proliferation and differentiation [18], facilitate T cell homing and migration [19–21] and ensure T cell survival [10, 18, 22, 23]. All three signals are required for the generation of efficient effector and memory T cell subsets from naïve T cells. If one of these three signals is not provided, T cell effector function is not optimal, or T cell tolerance is induced instead [16, 24, 25]. T cell dysfunction occurs similarly if the signal strength is not sufficient, or if these signals are delivered at the wrong time. Together, these mechanisms prevent unwanted T cell activation and immunopathology.

Upon efficient priming of naïve T cells, they differentiate into effector T cells. Effector T cells provide protection against pathogens through the secretion of cytolytic molecules, such as granzymes and perforin, and cytokines, such as interferon (IFN)-γ and tumor necrosis factor (TNF)-α [26–31]. Upon eradication of the invading pathogen or altered or tumorigenic cells, the pool of effector T cells shrinks, and a subset of T cells differentiates into memory T cells. Memory T cells provide long-lasting immunological memory to protect the host upon reinfection or cancers of the same origin [26]. While naïve CD8$^+$ T cells require all three signals for T cell activation, this is not the case for effector and memory CD8$^+$ T cells. This allows for the rapid production of effector molecules upon recognition of infected cells [26] and clearance of infections in early stages. In fact, TCR triggering alone is sufficient for the secretion of effector molecules by effector and memory CD8$^+$ T cells [29, 32].

However, while the production of these potent effector molecules by T cells is important for immune responses against cancer and invading pathogens, they also pose a risk for the host. Indeed, aberrant production of cytokines such as IFN-γ and TNF-α has been implicated in the development, maintenance or pathogenesis of diverse (immune) pathological diseases, including inflammatory bowel disease, multiple sclerosis and rheumatoid arthritis [33–39], and has also been shown to result in immunopathology in murine models [40–44].

Several (molecular) layers of regulation have been described that keep excessive T cell effector function in check. As described, for naïve T cell activation, three signals (TCR, co-stimulation and cytokine/chemokine signals) are required for optimal generation of effector and memory T cells. However, antigen-experienced T cells exhibit a lower activation threshold. For instance, besides TCR triggering, other signals can result in the induction of effector function in antigen-experienced T cells, including the sensing of pro-inflammatory molecules such as type I IFNs, interleukin (IL)-2, IL-12, IL-15 or IL-18 [45–47], or the triggering of damage- or pathogen-associated molecular pattern receptors such as toll-like receptors or cGAS/STING [48–51]. However, compared to TCR stimulation, non-TCR-mediated T cell activation is less optimal and leads only to a brief induction of cytokine production [29, 48]. Furthermore, in order to limit unbridled T cell activation after TCR triggering but also upon sensing pro-inflammatory cytokines, a transcriptional profile is engaged in T cells, resulting in the upregulation of inhibitory receptors [52–55]. Triggering of these inhibitory receptors, such as programmed cell death 1 (PD-1), inhibits the production of cytokines by T cells [56]. Conversely, blocking inhibitory receptor signaling enhances T cell effector function [57, 58] and is currently in use in the clinic for anticancer therapy. Another mechanism that is involved in maintaining T cell immune homeostasis is posttranscriptional regulation. Through the modulation of messenger ribonucleic acid (mRNA) splicing, localization, stability and translational rate, the posttranscriptional machinery ensures that the production of immune mediators occurs at the right time and at the right magnitude [30, 32, 33, 59, 60]. Here, the recent literature regarding the role of posttranscriptional regulation in maintaining immune homeostasis of antigen-experienced T cells is reviewed.

2 Posttranscriptional Events Governing T Cell Effector Function

In order to respond rapidly to invading pathogens, antigen-experienced T cells store effector molecules, such as granzymes [61], so they can be immediately released upon TCR triggering. Furthermore, also mRNA for inflammatory mediators, such as

IFN-γ, C-X-C motif chemokine ligand 10 (CXCL10, also known as IP-10), chemokine (C-C motif) ligand 4 (CCL4, also known as MIP-1β), CCL5 (also known as RANTES), and TNF-α, is stored [32, 59, 61–63]. However, these inflammatory proteins are not produced by resting antigen-experienced T cells. Only upon receiving a stimulatory signal, the protein of inflammatory molecules such as IFN-γ, CCL5 and TNF-α is produced [32, 59, 62]. Rather, the mRNAs of these inflammatory molecules are kept translationally inactive in a ready-to-go state, in order to be rapidly translated into protein upon encountering cognate antigen or receiving the appropriate stimulation [32, 59, 62]. In order to limit the excessive or unwanted production of inflammatory cytokines to the right place and the right time, posttranscriptional mechanisms ensure that cytokine protein is only formed from this preformed mRNA when T cells receive the right stimuli.

Posttranscriptional mechanisms are enforced through the binding of *trans* factors, such as noncoding RNAs or RNA-binding proteins (RBPs) to *cis* elements present in the 3′ or 5′ untranslated regions (UTRs) of mRNA. Several *cis* element and their *trans* binding partners have been identified to play a role in the regulation of cytokine production by T cells (recently reviewed in [60]). Here, the posttranscriptional events governed by micro RNAs (miRNAs) and adenylate-uridylate-rich elements (AREs) and their binding partners important for mediating immune homeostasis are reviewed.

2.1 miRNA-Mediated Posttranscriptional Regulation

miRNAs are ~22 nucleotide, single-stranded, RNA molecules that induce posttranscriptional events, regulating diverse cellular processes. Canonically, miRNAs are clustered in genetic regions, containing a few to several hundred different miRNA species. Often, these clusters are transcribed together as large transcripts containing several miRNAs. RNA polymerase II is the main polymerase associated with the transcription of miRNAs [64]. miRNA transcription results either in singular primary miRNA (pri-miRNA) transcripts containing one miRNA or a pri-miRNA transcript containing several miRNAs in the case of large polycistronic mRNAs. The actual ~22 nucleotide miRNA is located in the stem of a hairpin structure in pri-miRNAs. During miRNA processing (Fig. 2), first, pri-miRNAs are processed to single hairpins, also known as precursor miRNAs (pre-miRNAs), through interaction with a nuclear protein complex [65]. This nuclear protein complex, also known as microprocessor, consists of the RNAse III enzyme DROSHA, the double-stranded RNA-binding protein DiGeorge critical region 8 (DGCR8/Pasha) and other accessory factors, such as DEAD-Box helicase 5 (DDX5) and DDX17 [66]. Then, pre-miRNAs are exported out of the nucleus and into the cytoplasm. Here, processing of the pre-miRNA by the RNAse II enzyme DICER results in the formation of a double-stranded

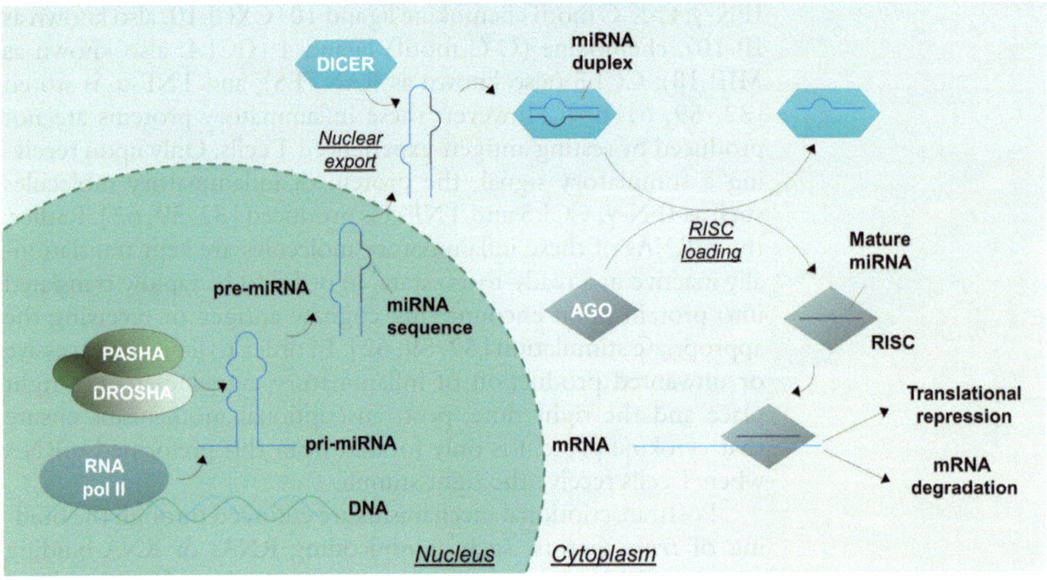

Fig. 2 miRNA processing and mode of action. miRNAs are transcribed in the nucleus from miRNA-containing genes and/or miRNA clusters by RNA polymerase II as pri-miRNAs. After processing to a stem-loop structure by the DROSHA/PASHA machinery, the now pre-miRNA is exported to the cytoplasm. Here, DICER cleaves the pre-miRNA to a miRNA duplex. The targeting strand is transferred to the AGO protein, forming the RISC complex. The miRNA-bearing RISC complex can now find its target mRNA species and exert its function, resulting in translational repression and/or mRNA degradation

RNA by cleavage of the loop [67, 68]. The now double-stranded RNA molecule consists of one mature miRNA, also known as the guide strand, and one redundant molecule, also known as the passenger strand. DICER associates with a member of the Argonaute family (AGO) and hands over the mature miRNA to AGO [69]. Subsequently, AGO and DICER dissociate, and the mature miRNA is loaded into the RNA-induced silencing complex (RISC) by AGO. The RISC complex is then guided by the miRNA to complementary sequences mainly present in 3′UTR transcripts of target mRNA molecules. At first, only translation is inhibited by the physical presence of the miRNA-loaded RISC, but later, miRNA binding also results in mRNA decay, halting or preventing protein production [70–72].

Part of the posttranscriptional events governing T cell cytokine production is mediated by noncoding RNAs such as miRNAs. Indeed, in T cells, miRNAs have been described to play a role in T cell development, differentiation and effector function. As such, miRNAs are instrumental in maintaining immune homeostasis. For instance, DICER-deficient T cells show impaired development compared to wild-type T cells and do not differentiate as efficiently [73]. DICER also plays a role in governing T cell effector function. It was shown that DICER-deficient murine T cells produce increased amounts of IFN-γ, IL-2 and TNF-α compared to wild-type T cells [73, 74]. Similarly, DICER-deficient human T cells produce more IFN-γ and TNF-α [75]. For TNF-α, it was shown that DICER is required for the degradation of *TNFA* mRNA in cell lines [76]. Furthermore, T cells from DICER knockout mice also expressed higher levels of *Prf1* (encoding the perforin protein) and produced more perforin upon stimulation [75]. Similarly, human DICER-deficient T cells contained more granzyme B and perforin protein compared to wild-type T cells [75].

As discussed, DICER is essential for miRNA generation. Thus, while these studies implicated a role for miRNAs in governing T cell effector function, a direct link was not shown. Indeed, the increased production of IFN-γ, IL-2 and TNF-α observed in DICER knockout mice was attributed to the regulation of mTOR complex components [74]. Potentially, this was due to deregulation of miR-15a/16, which have been shown to target mTOR [77]. miR-15a/16 deficiency also results in increased IFN-γ, IL-2 and TNF-α production by T cells [77].

Further research has tried to pinpoint which miRNAs are (directly) involved in the regulation of T cell effector molecules. For instance, the miR-29 family was shown to directly bind to and inhibit the translation of *Ifng* in murine T cells [78]. Several other miRNAs have since been found to be associated with an increased (e.g., miR-200a) or decreased (e.g., miR-181a and miR-322) IFN-γ production capacity by T cells [79]. However, whether these miRNAs directly regulate IFN-γ production through interaction with *IFNG* mRNA or whether they do so through the modulation of, for example, transcription factors remains unclear. miRNAs have also been implicated to regulate other T cell effector molecules. For instance, T cell–specific deletion of miR-31 resulted in increased expression of cytolytic molecules such as perforin and granzymes in knockout mice [80]. Similarly, also miR-139 and miR-342 were shown to be negative regulators of perforin [75]. An overview of which miRNAs have been implicated to regulate effector molecule production in T cells, either directly or indirectly, can be found in Table 1.

Together, these reports show the importance of miRNA production and miRNAs themselves in regulating the production of T cell effector molecules. While their role in steady state has not yet been elucidated, the potent regulation by miRNAs during T cell activation and differentiation alludes to a role for miRNAs to maintain immunological quiescence of inflammatory mediators in the absence of T cell stimuli. Indeed, deregulation of DICER or the miRNAs such as miR-29 or miR-31 results in deregulated production of effector molecules, potentially harming the host. Not surprisingly, miRNAs have been associated with autoimmune diseases [81]. Of note, direct binding of the discussed miRNAs to target transcripts was not shown on all occasions. Further research is

Table 1
miRNAs implicated to regulate T cell effector molecule production. – indicates negative regulation and + indicates positive regulation. Please note that not all references describe direct binding to target mRNAs

	Granzymes	IFN-γ	IL-2	Perforin	TNF-α	Reference(s)
miR-19b	+	+				[100]
miR-29		–				[78]
miR-31	–			–		[80]
miR-139				–		[75]
miR-149-3p		+	+		+	[101]
miR-181a		–				[79]
miR-181c			–			[102]
miR-155		+			+	[103, 104]
miR-342				–		[75]

required to fully understand miRNA-mediated posttranscriptional regulation of T cell effector molecules, both at steady state and during T cell activation.

2.2 ARE-Mediated Posttranscriptional Regulation

One RNA *cis* element that is of particular interest in the context of $CD8^+$ T cell immune homeostasis is the adenylate-uridylate-rich element, or ARE. AREs are located in the 3′UTR of mRNA transcripts [82, 83]. Several classes of AREs have been described based on their sequence identity [84–86]. Class I AREs consist of multimeric AUUUA pentamers [86]. The 3′UTR of the murine and the human *IFNG* gene contains five class I AREs [59]. Class II AREs consist of two overlapping UUAUUUA(U/A)(U/A) 9-mers and are encoded in the 3′UTR of the *IL2* and *TNFA* genes [85, 87]. Class III AREs are less well defined but are typically considered U-rich regions [85, 88]. Through computational modeling, it has been estimated that AREs are present in approximately 8–16% of mammalian genes [82, 83]. Interestingly, most cytokine-coding genes contain 3′UTR AREs [86, 89, 90].

A recent study investigating ARE-bearing transcripts in both resting memory human T cells and reactivated antigen-experienced human T cells showed an overall increase in ARE-bearing transcript in resting cells compared to transcripts with AREs [91]. However, when investigating secreted proteins only, mRNA levels declined when mRNA transcripts contained AREs, with a correlation between mRNA levels and the number of AREs present. In contrast, in reactivated antigen-experienced T cells, this correlation is reversed: in general, more mRNA is present from genes which contain multiple AREs compared to genes with only one or no AREs. Together, this highlights the important role of AREs in restraining the production of T cell effector molecules during homeostasis. Authors also attempted to correlate mRNA levels to protein levels, which indicated that ARE clusters are associated with the suppression of protein abundance [91]. However, as authors indicate, due to the low amount of matching mRNA and protein pairs, this result should be considered indicative only.

Also for the regulation of IFN-γ and TNF-α production, the 3′ UTR AREs are instrumental, especially to keep preexisting mRNA from being translated in the absence of T cell activation. Indeed, a murine model in which the 3′UTR AREs have been removed (ARE-Del) from the *Tnfa* gene showed that these mice have an impaired on/off regulation of TNF-α production. Ultimately, this leads to chronic inflammatory arthritis and Crohn's-like inflammatory bowel disease [33]. Similarly, deletion of the AREs from the *Ifng* gene resulted in systemic lupus erythematosus–like abnormalities [92] and loss of mesenchymal stromal cells [42] in mice. Furthermore, ex vivo incubation of ARE-Del T cells with Brefeldin A, a drug that blocks the secretion of proteins, showed that antigen-experienced T cells from *Ifng* ARE-Del mice spontaneously produced IFN-γ without the addition of exogenous peptide [59], indicating the importance of AREs in silencing IFN-γ production by T cells in the absence of a stimulatory signal. Of note, ARE-Del T cells also produce more *Ifng* mRNA and protein upon stimulation compared to wild-type T cells [58, 93]. This feature is conserved between mice and men, as more human ARE-Del T cells produce IFN-γ upon stimulation compared to wild-type T cells [30]. Interestingly, human ARE-Del T cells also were capable of sustaining IFN-γ production longer upon removal from stimuli. While IFN-γ production in wild-type cells typically ceased after 24 hours, *IFNG* ARE-Del T cells sustained production up to 72 hours [30]. Together, these reports show that ARE-mediated regulation is important for keeping IFN-γ production by T cells contained to the right place and the right time.

ARE-governed posttranscriptional events are mediated through the interaction with *trans* factors, for example, RBPs [60]. The binding of RBPs to AREs can result in the modulation of mRNA location, half-life and translational rate [59, 85, 94–96]. However, as described, several classes of AREs have been identified. Together with the surrounding sequence, these different classes or AREs facilitate the binding of different (sets of) RNA-binding proteins. Furthermore, the expression of RBPs is dependent on T cell activation status [97]. Thus, even though an RBP has been described as an ARE-binder, that alone does not determine whether it binds to a specific mRNA transcript. To further study this, an in-depth analysis of the RBP interactome of a specific mRNA species is required.

Recently, the RBP interactomes of *IFNG*, *TNFA* and *IL2* were explored [59, 96]. By making use of 4xS1m RNA aptamer loops fused to the 3′UTR of, for example, *IFNG*, authors identified 58 [96] and 80 [59] proteins that bound to the 3′UTR of *IFNG* in two independently performed experiments. The study of Salerno et al. focused on two of the 80 identified proteins, ZFP36 ring finger protein like 1 (ZFP36L1) and ZFP36L2 [59], members of the Tristetraprolin, or TTP, family of proteins, which have been described to bind to AREs [98] and have been shown to restrain antiviral $CD8^+$ T cell responses [99]. ZFP36L1 was only detectable upon T cell activation, while ZFP36L2 was present both in resting and in activated antigen-experienced T cells. Therefore, the authors hypothesized that ZFP36L2 must play a role in keeping the pre-formed *Ifng* mRNA in check [59]. In order to investigate this, Salerno et al. employed a T cell–specific ZFP36L2 knockout mouse model. Similar to the findings obtained from the ARE-Del mice, antigen-experienced ZFP36L2 knockout T cells produced IFN-γ protein in the absence of stimulation. However, while ARE-Del mice exhibit enhanced *Ifng* mRNA stability, ZFP36L2 mice do not, showing that ZFP36L2 does not alter *Ifng* mRNA half-life. Rather, binding of ZFP36L2 to the 3′UTR AREs of *Ifng* mRNA blocks the recruitment of *Ifng* mRNA into ribosomes, preventing translation in resting antigen-experienced T cells. Subsequent ZFP36L2 RNA immunoprecipitation experiments performed on lysates of resting antigen-experienced T cells also identified binding of ZFP36L2 to other T cell effector molecules, such as *Tnfa*, *Ccl3* and *Ccl4*. For TNF-α, authors showed that ZFP36L2 knockout mice also spontaneously produce TNF-α protein in the absence of stimulation, and that, similar to *Ifng*, *Tnfa* mRNA is more readily recruited to ribosomes in ZFP36L2 knockout mice. However, whether ZFP36L2 directly binds to the 3′UTR AREs present in the TNF-α gene remains to be determined. Furthermore, how ZFP36L2 is displaced upon T cell activation, freeing the cytokine mRNA for translation, is also still unclear.

In a follow-up study by Popović et al., in addition to *IFNG*, the interactome of *TNFA* and *IL2* was also explored [96]; 69 RBPs were identified as interacting with *TNFA* mRNA, while 82 RBPs interacted with *IL2*. Authors employed lysates from both resting and activated human T cells from the same donors and identified five strong interacting RBPs with mRNA of *IFNG*, *TNFA* and/or *IL2*: ataxin 2–like protein (ATXN2L), farnesoid X receptor 3 (FXR3), human antigen R (HuR), zinc finger CCCH-type containing antiviral 1 (ZC3HAV1) and, similarly to Salerno et al. [59], ZFP36L1 [96]. By using CRISPR/Cas9-mediated knockout, they showed that upon human T cell activation, ATXN2L, ZC3HAV1 and ZFP36L1 knockout results in higher percentages of IFN-γ^+, TNF-α^+ and IL-2^+ $CD8^+$ T cells, with each RBP influencing cytokine production at different stages of T cell activation. Conversely,

HuR knockout T cells produced less of these three key T cell cytokines. While FXR3 was found to be a strong interacting RBP, it did not have an effect on cytokine production in activated T cells. Authors next show that ZFP36L1 knockout T cells show increased mRNA levels and increased half-life of *IFNG*, *TNFA* and *IL2* mRNA. Interestingly, ATXN2L, HuR or ZC3HAV1 had no impact on mRNA levels or mRNA stability, hinting at other regulatory mechanisms. Potentially, these factors are involved in other steps of RNA regulation, such as splicing/localization (HuR is a nuclear export factor [95]), or influence mRNA translational rate. However, this remains to be determined.

Nonetheless, together, the reports of Salerno et al. and Popović et al. show the importance of ARE-binding RBPs in modulating T cell effector cytokines: ZFP36L2 silences translation of preformed *Ifng* mRNA in resting T cells, arming T cells with cytokine mRNA for rapid transcription upon reactivation, while ATXN2L, FXR3, ZC3HAV1 and ZFP36L1 keep excessive production of IFN-γ, TNF-α and IL-2 by activated T cells in check through posttranscriptional mechanisms to prevent unwanted immunopathology.

3 Conclusion and Outlook

Here, some of the posttranscriptional mechanisms ensuring antigen-experienced T cell quiescence in the absence of stimulation have been discussed, as well as some of the regulatory mechanisms that restrain excessive cytokine production of activated T cells. miRNAs and RNA *cis* elements such as AREs are instrumental in the regulation of T cell effector molecule production. In particular, the role of the AREs present in the 3′UTR of the *Ifng* and *Tnfa* genes is well defined. By functioning as binding sites for RBP proteins, in particular ZFP36L2, these AREs ensure that the cytokine mRNA is kept translationally inactive in resting T cells. Upon T cell activation, ZFP36L2 is displaced, resulting in rapid cytokine production. Together, this mechanism ensures that T cells do not produce cytotoxic molecules in the absence of stimulation but remain in a ready-to-go state to rapidly respond upon antigen reexposure. However, exactly how ZFP36L2 is displaced remains unclear. Furthermore, how other inflammatory molecules for which mRNAs are stored (e.g., CXCL10) is kept translationally inactive also requires further investigation.

On the other hand, RBPs such as ATXN2L, ZC3HAV1 and ZFP36L1 also play a role in maintaining homeostasis by restraining excessive production of inflammatory T cell cytokines IFN-γ, TNF-α and IL-2 [96]. Future studies investigating these mechanisms instrumental to maintaining host health could shed light on diseases where T cell effector function is deregulated (e.g., autoimmune disorders or cancer) and could provide clues as to how to steer towards a more balanced regulation of, for example, cytokine production in T cells.

References

1. Donskoy E, Goldschneider I (1992) Thrombocytopoiesis is maintained by blood-borne precursors throughout postnatal life. A study in parabiotic mice. J Immunol 148:1604–1612. https://doi.org/10.4049/jimmunol.148.6.1604
2. Spits H (2002) Development of αβ T cells in the human thymus. Nat Rev Immunol 2:760–772. https://doi.org/10.1038/nri913
3. De Smedt M, Reynvoet K, Kerre T et al (2002) Active form of notch imposes T cell fate in human progenitor cells. J Immunol 169:3021–3029. https://doi.org/10.4049/jimmunol.169.6.3021
4. Pallard C, Stegmann APA, Van Kleffens T et al (1999) Distinct roles of the phosphatidylinositol 3-kinase and STAT5 pathways in IL-7-mediated development of human thymocyte precursors. Immunity 10:525–535. https://doi.org/10.1016/S1074-7613(00)80052-7
5. Blom B, Verschuren MCM, Heemskerk MHM et al (1999) TCR gene rearrangements and expression of the pre-T cell receptor complex during human T-cell differentiation. Blood 93:3033–3043. https://doi.org/10.1182/blood.v93.9.3033
6. Palmer E (2003) Negative selection - clearing out the bad apples from the T-cell repertoire. Nat Rev Immunol 3:383–391. https://doi.org/10.1038/nri1085
7. Klein L, Kyewski B, Allen PM, Hogquist KA (2014) Positive and negative selection of the T cell repertoire: what thymocytes see (and don't see). Nat Rev Immunol 14:377–391. https://doi.org/10.1038/nri3667
8. Soerens AG, Künzli M, Quarnstrom CF et al (2023) Functional T cells are capable of supernumerary cell division and longevity. Nature. https://doi.org/10.1038/s41586-022-05626-9
9. Freen-van Heeren JJ, Opstelten R (2023) Age is just a number: long-lasting T-cell immunity can span several lifetimes. Immunol Cell Biol 101:374–376. https://doi.org/10.1111/imcb.12632
10. Kapsenberg ML (2003) Dendritic-cell control of pathogen-driven T-cell polarization. Nat Rev Immunol 3:984–993. https://doi.org/10.1038/nri1246
11. Acuto O, Michel F (2003) CD28-mediated co-stimulation: a quantitative support for TCR signalling. Nat Rev Immunol 3:939–951. https://doi.org/10.1038/nri1248
12. Hendriks J, Xiao Y, Borst J (2003) CD27 promotes survival of activated T cells and complements CD28 in generation and establishment of the effector T cell Pool. J Exp Med 198:1369–1380. https://doi.org/10.1084/jem.20030916
13. Nolte MA, van Olffen RW, van Gisbergen KPJM, van Lier RAW (2009) Timing and tuning of CD27 – CD70 interactions: the impact of signal strength in setting the balance between adaptive responses and immunopathology. Immunol Rev 229:216–231. https://doi.org/10.1111/j.1600-065x.2009.00774.x
14. Peperzak V, Veraar EAM, Xiao Y et al (2013) CD8+ T cells produce the chemokine CXCL10 in response to CD27/CD70 Costimulation to promote generation of the CD8+ effector T cell pool. J Immunol 191:3025–3036. https://doi.org/10.4049/jimmunol.1202222
15. Welten SPM, Redeker A, Franken KL et al (2013) CD27-CD70 costimulation controls T cell immunity during acute and persistent cytomegalovirus infection. J Virol 87:6851–6865. https://doi.org/10.1128/JVI.03305-12
16. Curtsinger JM, Lins DC, Mescher MF (2003) Signal 3 determines tolerance versus full activation of naive CD8 T cells. J Exp Med 197:1141–1151. https://doi.org/10.1084/jem.20021910
17. Curtsinger JM, Mescher MF (2010) Inflammatory cytokines as a third signal for T cell activation. Curr Opin Immunol 22:333–340. https://doi.org/10.1016/j.coi.2010.02.013
18. Hwang JR, Byeon Y, Kim D, Park SG (2020) Recent insights of T cell receptor-mediated signaling pathways for T cell activation and development. Exp Mol Med 52:750–761. https://doi.org/10.1038/s12276-020-0435-8
19. Marelli-Berg FM, Okkenhaug K, Mirenda V (2007) A two-signal model for T cell trafficking. Trends Immunol 28:267–273. https://doi.org/10.1016/j.it.2007.04.008
20. Gregor CE, Foeng J, Comerford I, McColl SR (2017) Chemokine-driven CD4+ T cell homing: new concepts and recent advances. Adv Immunol 135:119–181. https://doi.org/10.1016/bs.ai.2017.03.001
21. Martens R, Permanyer M, Werth K et al (2020) Efficient homing of T cells via afferent lymphatics requires mechanical arrest and integrin-supported chemokine guidance. Nat Commun 11. https://doi.org/10.1038/s41467-020-14921-w

22. Boise LH, Minn AJ, Noel PJ et al (1995) CD28 costimulation can promote T cell survival by enhancing the expression of Bcl-XL. Immunity 3:87–98. https://doi.org/10.1016/1074-7613(95)90161-2
23. Haring JS, Badovinac VP, Harty JT (2006) Inflaming the CD8+ T cell response. Immunity 25:19–29. https://doi.org/10.1016/j.immuni.2006.07.001
24. Li Y, Li XC, Zheng XX et al (1999) Blocking both signal 1 and signal 2 of T-cell activation prevents apoptosis of alloreactive t cells and induction of peripheral allograft tolerance. Nat Med 5:1298–1302. https://doi.org/10.1038/15256
25. Sckisel GD, Bouchlaka MN, Monjazeb AM et al (2015) Out-of-sequence signal 3 paralyzes primary CD4+ T-cell-dependent immunity. Immunity 43:240–250. https://doi.org/10.1016/j.immuni.2015.06.023
26. Pennock ND, White JT, Cross EW et al (2013) T cell responses: naïve to memory and everything in between. Adv Psyiol Educ 37:273–283. https://doi.org/10.1152/advan.00066.2013
27. Voskoboinik I, Whisstock JC, Trapani JA (2015) Perforin and granzymes: function, dysfunction and human pathology. Nat Rev Immunol 15:388–400. https://doi.org/10.1038/nri3839
28. Golubovskaya V, Wu L (2016) Different subsets of T cells, memory, effector functions, and CAR-T immunotherapy. Cancers (Basel) 8. https://doi.org/10.3390/cancers8030036
29. Salerno F, Freen-van Heeren JJ, Guislain A et al (2019) Costimulation through TLR2 drives polyfunctional CD8$^+$ T cell responses. J Immunol 202:714–723. https://doi.org/10.4049/jimmunol.1801026
30. Freen-van Heeren JJ, Popović B, Guislain A, Wolkers MC (2020) Human T cells employ conserved AU-rich elements to fine-tune IFN-γ production. Eur J Immunol 50:949–958. https://doi.org/10.1002/eji.201948458
31. Nicolet BP, Guislain A, van Alphen FPJ et al (2020) CD29 marks superior cytotoxic human CD8+ T cells. Proc Natl Acad Sci 117:6686–6696. https://doi.org/10.1101/562512
32. Salerno F, Paolini NA, Stark R et al (2017) Distinct PKC-mediated posttranscriptional events set cytokine production kinetics in CD8+ T cells. Proc Natl Acad Sci 114:9677–9682. https://doi.org/10.1073/pnas.1704227114
33. Kontoyiannis D, Pasparakis M, Pizarro TT et al (1999) Impaired on/off regulation of TNF biosynthesis in mice lacking TNF AU-rich elements: implications for joint and gut-associated immunopathologies. Immunity 10:387–398. https://doi.org/10.1016/S1074-7613(00)80038-2
34. Mori L, Iselin S, De Libero G, Lesslauer W (1996) Attenuation of collagen-induced arthritis in 55-kDa TNF receptor type 1 (TNFR1)-IgG1-treated and TNFR1-deficient mice. J Immunol 157:3178–3182
35. Aardoom MA, Veereman G, De Ridder L (2019) A review on the use of anti-TNF in children and adolescents with inflammatory bowel disease. Int J Mol Sci 20:2529. https://doi.org/10.3390/ijms20102529
36. Monaco C, Nanchahal J, Taylor P, Feldmann M (2014) Anti-TNF therapy: past, present and future. Int Immunol 27:55–62. https://doi.org/10.1093/intimm/dxu102
37. Bisping G, Lugering N, Lutke-Brintrup S et al (2001) Patients with inflammatory bowel disease (IBD) reveal increased induction capacity of intracellular interferon-gamma (IFN-g) in peripheral CD8+ lymphocytes co-cultured with intestinal epithelial cells. Clin Exp Immunol 123:15–22. https://doi.org/10.1046/j.1365-2249.2001.01443.x
38. Ito R, Shin-Ya M, Kishida T et al (2006) Interferon-γ is causatively involved in experimental inflammatory bowel disease in mice. Clin Exp Immunol 146:330–338. https://doi.org/10.1111/j.1365-2249.2006.03214.x
39. Rafa H, Amri M, Saoula H et al (2010) Involvement of interferon-γ in bowel disease pathogenesis by nitric oxide pathway: a study in Algerian patients. J Interf Cytokine Res 30:691–697. https://doi.org/10.1089/jir.2010.0012
40. Yoshinaga M, Takeuchi O (2019) RNA binding proteins in the control of autoimmune diseases. Immunol Med 42:53–64. https://doi.org/10.1080/25785826.2019.1655192
41. De Bruin AM, Voermans C, Nolte MA (2014) Impact of interferon-γ on hematopoiesis. Blood 124:2479–2486. https://doi.org/10.1182/blood-2014-04-568451
42. Goedhart M, Cornelissen AS, Kuijk C et al (2018) Interferon-gamma impairs maintenance and alters hematopoietic support of bone marrow mesenchymal stromal cells. Stem Cells Dev 27:579–589. https://doi.org/10.1089/scd.2017.0196
43. de Bruin AM, Libregts SF, Valkhof M et al (2012) IFN-γ induces monopoiesis and

inhibits neutrophil development during inflammation. Blood 119:1543–1555. https://doi.org/10.1182/blood-2011-07-367706.An
44. Libregts SF, Gutie L, De Bruin AM et al (2011) Chronic IFN-γ production in mice induces anemia by reducing erythrocyte life span and inhibiting erythropoiesis through an IRF-1/PU.1 axis. Blood 118:2578–2589. https://doi.org/10.1182/blood-2010-10-315218
45. Kohlmeier JE, Cookenham T, Roberts AD et al (2010) Type I interferons regulate cytolytic activity of memory CD8+ T cells in the lung airways during respiratory virus challenge. Immunity 33:96–105. https://doi.org/10.1016/j.immuni.2010.06.016.Type
46. Berg RE, Cordes CJ, Forman J (2002) Contribution of CD8+ T cells to innate immunity: IFN-γ secretion induced by IL-12 and IL-18. Eur J Immunol 32:2807–2816. https://doi.org/10.1002/1521-4141(2002010)32:10%3C2807::aid-immu2807%3E3.0.co;2-0
47. Berg RE, Crossley E, Murray S, Forman J (2003) Memory CD8+ T cells provide innate immune protection against listeria monocytogenes in the absence of cognate antigen. J Exp Med 198:1583–1593. https://doi.org/10.1084/jem.20031051
48. Salerno F, Guislain A, Cansever D, Wolkers MC (2016) TLR-mediated innate production of IFN-γ by CD8+ T cells is independent of glycolysis. J Immunol 196:3695–3705. https://doi.org/10.4049/jimmunol.1501997
49. Li Q, Yan Y, Liu J et al (2019) Toll-like receptor 7 activation enhances CD8+ T cell effector functions by promoting cellular glycolysis. Front Immunol 10. https://doi.org/10.3389/fimmu.2019.02191
50. Zhang E, Ma Z, Li Q et al (2019) TLR2 stimulation increases cellular metabolism in CD8 + T cells and thereby enhances CD8 + T cell activation, function, and antiviral activity. J Immunol 203:2872–2288. https://doi.org/10.4049/jimmunol.1900065
51. Freen-van Heeren JJ (2021) TLR-2/7-mediated T cell activation: an innate potential to augment CD8^{+} T cell cytokine production. Scand J Immunol 93:e13019. https://doi.org/10.1111/sji.13019
52. Agata Y, Kawasaki A, Nishimura H et al (1996) Expression of the PD-1 antigen on the surface of stimulated mouse T and B lymphocytes. Int Immunol 8:765–772. https://doi.org/10.1093/intimm/8.5.765
53. Linsley PS, Bradshaw J, Greene JA et al (1996) Intracellular trafficking of CTLA-4 and focal localization towards sites of TCR engagement. Immunity 4:535–543. https://doi.org/10.1016/S1074-7613(00)80480-X
54. Kinter AL, Godbout EJ, McNally JP et al (2008) The common γ-chain cytokines IL-2, IL-7, IL-15, and IL-21 induce the expression of programmed death-1 and its ligands. J Immunol 181:6738–6746. https://doi.org/10.4049/jimmunol.181.10.6738
55. Terawaki S, Chikuma S, Shibayama S et al (2011) IFN-α directly promotes programmed cell death-1 transcription and limits the duration of T cell-mediated immunity. J Immunol 186:2772–2779. https://doi.org/10.4049/jimmunol.1003208
56. Wei F, Zhong S, Ma Z et al (2013) Strength of PD-1 signaling differentially affects T-cell effector functions. Proc Natl Acad Sci U S A 110:2–11. https://doi.org/10.1073/pnas.1305394110
57. Zhang Y, Huang S, Gong D et al (2010) Programmed death-1 upregulation is correlated with dysfunction of tumor-infiltrating CD8+ T lymphocytes in human non-small cell lung cancer. Cell Mol Immunol 7:389–395. https://doi.org/10.1038/cmi.2010.28
58. Salerno F, Guislain A, Freen-van Heeren JJ et al (2018) Critical role of post-transcriptional regulation for IFN-γ in tumor-infiltrating T cells. OncoImmunology 8. https://doi.org/10.1080/2162402X.2018.1532762
59. Salerno F, Engels S, van den Biggelaar M et al (2018) Translational repression of pre-formed cytokine-encoding mRNA prevents chronic activation of memory T cells. Nat Immunol 19:828–837. https://doi.org/10.1038/s41590-018-0155-6
60. Freen-van Heeren JJ (2021) Post-transcriptional control of T-cell cytokine production: implications for cancer therapy. Immunology 164:57–72. https://doi.org/10.1111/imm.13339
61. Kaech SM, Hemby S, Kersh E, Ahmed R (2002) Molecular and functional profiling of memory CD8 T cell differentiation. Cell 111:837–851. https://doi.org/10.1016/S0092-8674(02)01139-X
62. Swanson BJ, Murakami M, Mitchell TC et al (2002) RANTES production by memory phenotype T cells is controlled by a posttranscriptional, TCR-dependent process. Immunity 17:605–615. https://doi.org/10.1016/S1074-7613(02)00456-9

63. Oja AE, Piet B, Helbig C et al (2018) Trigger-happy resident memory CD4+ T cells inhabit the human lungs. Mucosal Immunol 11:654–667. https://doi.org/10.1038/mi.2017.94
64. Lee Y, Kim M, Han J et al (2004) MicroRNA genes are transcribed by RNA polymerase II. EMBO J 23:4051–4060. https://doi.org/10.1038/sj.emboj.7600385
65. Lee Y, Ahn C, Han J et al (2003) The nuclear RNase III Drosha initiates microRNA processing. Nature 425:415–419. https://doi.org/10.1038/nature01957
66. Treiber T, Treiber N, Meister G (2019) Regulation of microRNA biogenesis and its crosstalk with other cellular pathways. Nat Rev Mol Cell Biol 20:5–20. https://doi.org/10.1038/s41580-018-0059-1
67. Grishok A, Pasquinelli AE, Conte D et al (2001) Genes and mechanisms related to RNA interference regulate expression of the small temporal RNAs that control C. elegans developmental timing. Cell 106:23–34. https://doi.org/10.1016/S0092-8674(01)00431-7
68. Ketting RF, Fischer SEJ, Bernstein E et al (2001) Dicer functions in RNA interference and in synthesis of small RNA involved in developmental timing in C. elegans. Genes Dev 15:2654–2659. https://doi.org/10.1101/gad.927801
69. Kobayashi H, Tomari Y (2016) RISC assembly: coordination between small RNAs and Argonaute proteins. Biochim Biophys Acta Gene Regul Mech 1859:71–81. https://doi.org/10.1016/j.bbagrm.2015.08.007
70. Bazzini AA, Lee MT, Giraldez AJ (2012) Ribosome profiling shows that miR-430 reduces translation before causing mRNA decay in Zebrafish. Science 80(336):233–237. https://doi.org/10.1126/science.1215704
71. Béthune J, Artus-Revel CG, Filipowicz W (2012) Kinetic analysis reveals successive steps leading to miRNA-mediated silencing in mammalian cells. EMBO Rep 13:716–723. https://doi.org/10.1038/embor.2012.82
72. Djuranovic S, Nahvi A, Green R (2012) miRNA-mediated gene silencing by translational repression followed by mRNA deadenylation and decay. Science 80(336):237–240. https://doi.org/10.1126/science.1215691
73. Muljo SA, Ansel KM, Kanellopoulou C et al (2005) Aberrant T cell differentiation in the absence of Dicer. J Exp Med 202:261–269. https://doi.org/10.1084/jem.20050678
74. Marcais A, Blevins R, Graumann J et al (2014) microRNA-mediated regulation of mTOR complex components facilitates discrimination between activation and anergy in CD4 T cells. J Exp Med 211:2281–2295. https://doi.org/10.1084/jem.20132059
75. Trifari S, Pipkin ME, Bandukwala HS et al (2013) MicroRNA-directed program of cytotoxic CD8+ T-cell differentiation. Proc Natl Acad Sci 110:18608–18613. https://doi.org/10.1073/pnas.1317191110
76. Jing Q, Huang S, Guth S et al (2005) Involvement of MicroRNA in AU-rich element-mediated mRNA instability. Cell 120:623–634. https://doi.org/10.1016/j.cell.2004.12.038
77. Yang J, Liu R, Deng Y et al (2017) MiR-15a/16 deficiency enhances anti-tumor immunity of glioma-infiltrating CD8+ T cells through targeting mTOR. Int J Cancer 141:2082–2092. https://doi.org/10.1002/ijc.30912
78. Ma F, Xu S, Liu X et al (2011) The microRNA miR-29 controls innate and adaptive immune responses to intracellular bacterial infection by targeting interferon-γ. Nat Immunol 12:861–869. https://doi.org/10.1038/ni.2073
79. Amado T, Amorim A, Enguita FJ et al (2020) MicroRNA-181a regulates IFN-γ expression in effector CD8+ T cell differentiation. J Mol Med 98:309–320. https://doi.org/10.1007/s00109-019-01865-y
80. Moffett HF, Cartwright ANR, Kim H et al (2017) The microRNA miR-31 inhibits CD8+ T cell function in chronic viral infection. Nat Immunol 18:791–799. https://doi.org/10.1038/ni.3755.The
81. Salvi V, Gianello V, Tiberio L et al (2019) Cytokine targeting by miRNAs in autoimmune diseases. Front Immunol 10. https://doi.org/10.3389/fimmu.2019.00015
82. Bakheet T, Frevel M, Williams BR et al (2001) ARED: human AU-rich element-containing mRNA database reveals an unexpectedly diverse functional repertoire of encoded proteins. Nucleic Acids Res 29:246–254. https://doi.org/10.1093/nar/29.1.246
83. Gruber AR, Fallmann J, Kratochvill F et al (2011) AREsite: a database for the comprehensive investigation of AU-rich elements. Nucleic Acids Res 39:D66–D69. https://doi.org/10.1093/nar/gkq990
84. Xu N, Chen C-YA, Shyu A-B (1997) Modulation of the fate of cytoplasmic mRNA by AU-rich elements: key sequence features controlling mRNA deadenylation and decay

85. Barreau C, Paillard L, Osborne HB (2005) AU-rich elements and associated factors: are there unifying principles? Nucleic Acids Res 33:7138–7150. https://doi.org/10.1093/nar/gki1012
86. Beisang D, Bohjanen P (2012) Perspectives on the ARE as it turns 25 years old. Wiley Interdiscip Rev RNA 3:719–731. https://doi.org/10.1002/wrna.1125
87. Gueydan C, Droogmans L, Chalon P et al (1999) Identification of TIAR as a protein binding to the translational regulatory AU-rich element of tumor necrosis factor α mRNA. J Biol Chem 274:2322–2326. https://doi.org/10.1074/jbc.274.4.2322
88. Peng SS, Chen CY, Shyu AB (1996) Functional characterization of a non-AUUUA AU-rich element from the c-Jun proto-oncogene mRNA: evidence for a novel class of AU-rich elements. Mol Cell Biol 16:1490–1499. https://doi.org/10.1128/mcb.16.4.1490
89. Salerno F, Wolkers MC (2015) T-cells require post-transcriptional regulation for accurate immune responses. Biochem Soc Trans 43:1201–1207. https://doi.org/10.1042/BST20150154
90. Vlasova-St. Louis I, Bohjanen PR (2014) Post-transcriptional regulation of cytokine signaling by AU-rich and GU-rich elements. J Interf Cytokine Res 34:233–241. https://doi.org/10.1089/jir.2013.0108
91. Nicolet BP, Wolkers MC (2022) The relationship of mRNA with protein expression in CD8+ T cells associates with gene class and gene characteristics. PLoS One 17:4–6. https://doi.org/10.1371/journal.pone.0276294
92. Hodge DL, Berthet C, Coppola V et al (2014) IFN-gamma AU-rich element removal promotes chronic IFN-gamma expression and autoimmunity in mice. J Autoimmun 53:33–45. https://doi.org/10.1016/j.jaut.2014.02.003
93. Freen-van Heeren JJ, Nicolet BP, Wolkers MC (2020) Combined single-cell measurement of cytokine mRNA and protein in immune cells. Methods Mol Biol 2108:259–271. https://doi.org/10.1007/978-1-0716-0247-8_22
94. Shim J, Lim H, Yates JR, Karin M (2002) Nuclear export of NF90 is required for interleukin-2 mRNA stabilization. Mol Cell 10:1331–1344. https://doi.org/10.1016/S1097-2765(02)00730-X
95. Wang JG, Collinge M, Ramgolam V et al (2006) LFA-1-dependent HuR nuclear export and cytokine mRNA stabilization in T cell activation. J Immunol 176:2105–2113. https://doi.org/10.4049/jimmunol.176.4.2105
96. Popovic B, Guislain A, Engels S et al (2023) Time-dependent regulation of cytokine production by RNA binding proteins defines T cell effector function. Cell Rep 42:112419. https://doi.org/10.1016/j.celrep.2023.112419
97. Zandhuis ND, Nicolet BP, Wolkers MC (2021) RNA-binding protein expression alters upon differentiation of human B cells and T cells. Front Immunol 12:1–16. https://doi.org/10.3389/fimmu.2021.717324
98. Hodson DJ, Janas ML, Galloway A et al (2010) Deletion of the RNA-binding proteins ZFP36L1 and ZFP36L2 leads to perturbed thymic development and T lymphoblastic leukemia. Nat Immunol 11:717–724. https://doi.org/10.1038/ni.1901
99. Petkau G, Mitchell TJ, Chakraborty K et al (2022) The timing of differentiation and potency of CD8 effector function is set by RNA binding proteins. Nat Commun 13:1–16. https://doi.org/10.1038/s41467-022-29979-x
100. Yin LB, Song CB, Zheng JF et al (2019) Elevated expression of miR-19b enhances CD8+T cell function by targeting PTEN in HIV infected long term non-progressors with sustained viral suppression. Front Immunol 10:1–12. https://doi.org/10.3389/fimmu.2018.03140
101. Zhang M, Gao D, Shi Y et al (2019) miR-149-3p reverses CD8+ T-cell exhaustion by reducing inhibitory receptors and promoting cytokine secretion in breast cancer cells. Open Biol 9:190061. https://doi.org/10.1098/rsob.190061
102. Xue Q, Guo ZY, Li W et al (2011) Human activated CD4+ T lymphocytes increase IL-2 expression by downregulating microRNA-181c. Mol Immunol 48:592–599. https://doi.org/10.1016/j.molimm.2010.10.021
103. Ji Y, Wrzesinski C, Yu Z et al (2015) MiR-155 augments CD8+ T-cell antitumor activity in lymphoreplete hosts by enhancing responsiveness to homeostatic γc cytokines. Proc Natl Acad Sci U S A 112:476–481. https://doi.org/10.1073/pnas.1422916112
104. Monnot GC, Martinez-Usatorre A, Lanitis E et al (2020) miR-155 overexpression in OT-1 CD8+ T cells improves anti-tumor activity against low-affinity tumor antigen. Mol Ther Oncolytics 16:111–123. https://doi.org/10.1016/j.omto.2019.12.008

Chapter 5

A Robust In Vitro Culture Model and Generation of Memory Monocytes

Jing Wang, Shuo Geng, Yajun Wu, and Liwu Li

Abstract

Innate monocytes can be trained or reprogrammed to adopt distinct memory states, such as low-grade inflammation and immune exhaustion, bearing fundamental relevance to the pathogenesis of both acute diseases such as sepsis as well as chronic diseases such as atherosclerosis. Therefore, it is critically important to develop a regimen for generating memory monocytes in vitro in order to better define key monocyte memory states with diverse potentials for proliferation, differentiation, and activation, as well as underlying mechanisms. Here, we describe an efficient in vitro system to propagate a large number of highly purified murine memory monocytes through sustaining bone marrow–derived monocytes with macrophage colony-stimulating factor (M-CSF, 10 ng/mL)-containing medium, together with other polarization agents such as lipopolysaccharide (LPS) for a 5-day period. This method can yield high-purity monocytes, capable of exhibiting dynamic memory behaviors upon training with various polarizing agents.

Key words Model system, Immune memory generation, Bone marrow culture, Priming, Exhaustion

1 Introduction

In the symbiosis with bacteria, mammals are intimately impacted by the bacterial endotoxin lipopolysaccharide (LPS) in their circulation [1, 2]. Depending on diverse physiological and pathological conditions, circulating LPS can range from super-low dose (subclinical endotoxemia associated with chronic inflammatory diseases) to high dose associated with acute sepsis. Mechanistically, prolonged exposure to subclinical dose LPS polarizes monocytes into a non-resolving low-grade inflammatory state which contributes to the pathogenesis of atherosclerosis and impaired wound repair [3, 4]. On the other hand, prolonged exposure to high-dose LPS drives the generation of an exhausted monocyte memory state reflected in immune suppression and pathogenic inflammation

Jing Wang and Shuo Geng have equally contributed to this chapter.

Vijay Kumar (ed.), *Immune Homeostasis: Methods and Protocols*, Methods in Molecular Biology, vol. 2782,
https://doi.org/10.1007/978-1-0716-3754-8_5,

collectively leading to septic shock [5, 6]. Exhausted septic monocytes are less mature and express high levels of pathogenic inflammatory markers such as CD38 as well as high levels of immune suppressive markers such as PD-L1 [5]. Septic monocytes also exhibit compromised expression of immune-enhancing mediators such as CD86, collectively leading to pathogenic inflammation and immune suppression [5, 7]. Despite their emerging relevance, robust in vitro systems in clearly capturing monocyte memory generation and underlying mechanisms are limited.

Here, we describe a robust and efficient in vitro culture system for sustaining the generation of murine memory monocytes [5]. All bone marrow cells are cultured with macrophage colony-stimulating factor (M-CSF)-containing medium at 37 °C in a humidified 5% CO_2 atmosphere for a 5-day period, which will preferentially allow the sustained propagation of monocyte population adopting dynamic behaviors, in the presence of additional polarizing agents such as LPS. As we reported previously, bone marrow cells cultured under this condition were monocyte-like, loosely adherent and did not express the mature macrophage marker CD71 [8]. Using this method, we can generate more than 95% pure monocytes. Single-cell RNA-sequencing (scRNAseq) analyses reveal the generation of monocytes with potentials of proliferation, and/or subsequent polarization into anti-inflammatory, pro-inflammatory, or exhausted states depending upon the nature and strength of added stimulants [5, 9].

2 Materials

Prepare all cell culture solutions using sterile regent or filtered with a 0.22 μm membrane filter. Prepare and store all cell culture solutions at 4 °C (stock solution should be stored at −20 °C). Follow all waste disposal regulations when disposing waste materials.

2.1 Bone Marrow Cell Isolation and Culture

1. 6-well cell culture plates.
2. 50 mL conical tubes.
3. 15 mL conical tubes.
4. Anesthetics (isoflurane).
5. 1 mL syringe (with 25-gauge needle).
6. 70 μm cell strainers.
7. Red blood cell (RBC) lysis buffer (ammonium-chloride-potassium lysing buffer).
8. Phosphate-buffered saline (PBS).
9. Flow cytometry staining (FACS) buffer: Hanks' balanced salt solution containing 2% fetal bovine serum (FBS).

10. Complete RPMI-1640 (Roswell Park Memorial Institute) Medium: RPMI-1640 medium supplemented with 10% FBS, 1% penicillin-streptomycin (10,000 U/mL), and 1% L-glutamine.
11. M-CSF-containing medium: Complete RPMI-1640 medium supplemented with 10 ng/mL M-CSF. (Medium should be freshly prepared before cell culture and each medium change.)
12. 1 mg/mL LPS from *Escherichia coli* O111:B4 (Sigma, #L2630): LPS powder is dissolved in 10 mL of PBS, and then LPS stock solution is aliquoted into 1 mL per Eppendorf tube and stored at −20 °C. For using a concentration of 10 μg/mL LPS: Dilute 10 μL of 1 mg/mL LPS into 990 μL of complete RPMI-1640 medium and stock at 4 °C before use. (It is better to prepare fresh LPS working solution before each experiment.)

2.2 Flow Cytometry Test

1. Fc blocker: Purified rat anti-mouse CD16/CD32 (BD Biosciences, #533142).
2. Propidium iodide (PI): 1.0 mg/mL solution in water.
3. APC (Allophycocyanin)/Cyanine7 anti-mouse/human CD11b antibody (Biolegend, # 101226).
4. PE (Phycoerythrin)/Cyanine7 anti-mouse Ly-6C antibody (Biolegend, # 128018).
5. FITC (Fluorescein isothiocyanate) anti-mouse CD86 antibody (Biolegend, # 105005).
6. APC anti-mouse CD38 antibody (Biolegend, # 102712).
7. PE anti-mouse CD274 (B7-H1, PD-L1) antibody (Biolegend, # 124308).

3 Methods

3.1 Bone Marrow Harvest

1. Euthanize mice using isoflurane, followed by cervical dislocation. Under sterile conditions (*see* **Notes 1** and **4**), isolate the tibia and femur (both left and right) and cut at the joints from mice. Place tibia and femur into a sterile 6-well cell culture plate (with 3 mL of ice-cold RPMI-1640).
2. Use small scissors and forceps to remove the muscle and fibrous tissues from the tibia and femur (*see* **Note 5**). If the muscles are still intact, a Kimwipe saturated with 70% ethanol can be used to wipe the bones to remove any excess muscle fibers (*see* **Note 2**).
3. Sterilize the tibia and femur by dipping in 70% ethanol, and then wash the bones 3 times with ice-cold RPMI-1640.

4. Transfer the tibia and femur into a new well (6-well cell culture dish, with 3 mL of ice-cold RPMI-1640). Cut both tibia and femur ends with a scalpel or sharp scissors so that both ends are clean.
5. Attach a 25-gauge needle to a 1 mL syringe. Draw up 1 mL of RPMI-1640 medium from 3 mL RPMI-1640-containing dish. Using forceps to hold the bone above the cell culture dish, carefully flush out the marrow into the cell culture dish using the syringe containing 1 mL of RPMI-1640. Repeat flushing as necessary until the bone is white, which indicates that all the marrow has been removed. Repeat this step for all bones.
6. Collect cells into 15 mL conical tubes and centrifuge cells (400 g, 6 min, 4 °C).
7. Remove supernatant and resuspend cells in 1 mL of cold RBC lysis buffer, mix well, and incubate on ice for 5 min (to lyse RBC).
8. After 5 min, add 9 mL of cold PBS to stop the reaction. Centrifuge cells (400 g, 6 min, 4 °C). Remove supernatant and resuspend cells in 5 mL of pre-warmed M-CSF-containing medium (*see* **Note 3**).
9. Place a 70 μm cell strainer on top of a new 15 mL conical tube, held upright in a tube rack. Use a 1 mL pipette to filter all resuspended cells into a new 15 mL conical tube to remove cell clots.
10. Count cell numbers and adjust concentration to 10^6 cells/mL (using M-CSF-containing medium to dilute cells).

3.2 Culture with M-CSF-Containing Medium

1. After adjusting the concentration of bone marrow cells, culture cells in 6-well plates. Add 3 mL of cell suspension to each well (3×10^6 cells/well) for 5 days. To generate exhausted memory monocytes, high-dose LPS (100 ng/mL) is added into the cell culture medium for 5 days (*see* **Note 8**).
2. Add 30 μL of 10 μg/mL LPS to each cell culture wells (*see* **Note 9**). For the control group, add 30 μL of PBS. Evenly distribute plate and culture cells at 37 °C in a humidified 5% CO_2 atmosphere (Day 0).
3. After 2 days of cell culture (Day 2), take out the cell culture plate from the incubator and gently distribute into the cell culture hood. Remove 1.5 mL of cell culture supernatant from each well and then add 3.5 mL of fresh pre-warmed M-CSF supplemented medium (*see* **Note 6**). For the memory monocyte culture group, add 50 μL of 10 μg/mL LPS to each well. Add 50 μL of PBS to each well as control.
4. After another 2 days of cell culture (Day 4), take out the cell culture plate and gently distribute. Remove 2.5 mL of cell

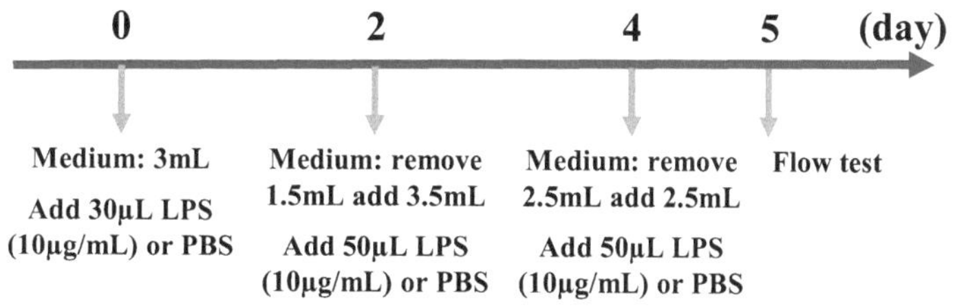

Fig. 1 Timeline of cell culture. Cells were treated with either PBS (control) or high-dose LPS (final concentration 100 ng/mL) for 5 days at 37 °C in a humidified 5% CO_2 atmosphere. Fresh M-CSF-containing medium and treatments (PBS or LPS) were added every 2 days. For the short-time high-dose LPS (100 ng/mL) group, LPS was added on day 5 (50 μL of 10 μg/mL LPS for each well)

culture supernatant from each well and then add 2.5 mL of fresh pre-warmed M-CSF supplemented medium. For the memory monocyte culture group, add 50 μL of 10 μg/mL LPS to each well. Add 50 μL of PBS to each well as control.

5. Cultured cells can be tested by relevant analyses such as flow cytometry (Fig. 1).

3.3 Analyses

1. Gently aspirate cell culture medium from a 6-well cell culture plate. Using 1 mL of cold PBS, wash each well three times, then add 1 mL of PBS to each well, and put the plate on ice for 15–20 min.
2. After 15–20 min incubation, use a cell scraper to gently scrape monocytes from the cell culture plates (*see* **Note** 7). Collect cells into a 15 mL conical tube and spin down cells (400 g, 6 min, 4 °C).
3. Remove supernatant and resuspend the cells with blocker buffer (1:100 Fc blocker in FACS buffer. Use 25 μL for each well of cells). Incubate on ice for 20 min.
4. Add 25 μL antibody cocktail (1:100 diluted APC/Cyanine7 anti-mouse/human CD11b and PE/Cyanine7 anti-mouse Ly-6C, demonstrating >99% of cells being CD11b positive monocytes (Fig. 2). Using this approach, we also reported that cultured monocytes can adopt rich patho-physiologically relevant behaviors such as proliferation, differentiation, and activation [5, 8]. Just as another example, cells can be stained with other intended activation markers such as FITC anti-mouse CD86, APC anti-mouse CD38, and PE anti-mouse PD-L1) to each blocked sample. Thus, the final dilution of each antibody is 1:200.
5. After 30 min 4 °C incubation, wash cells with 200 μL FACS buffer twice (400 g, 6 min, 4 °C). Resuspend cells in FACS buffer containing PI (1:2000).

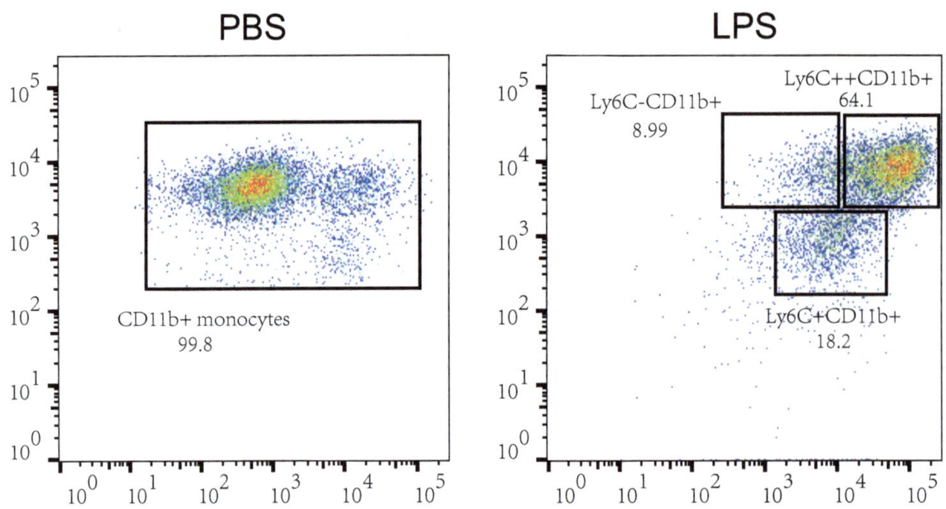

Fig. 2 Flow illustration of monocytes after 5 days of LPS stimulation. Bone marrow cells are cultured with M-CSF containing medium for 5 days. Cells are treated with either PBS or high-dose LPS (100 ng/mL) for 5 days in incubator, and > 99% of cells are CD11b positive monocytes after 5-day culture (Left panels). Ly6C$^-$ CD11b$^+$, Ly6C$^+$ CD11b$^+$, and Ly6C^{++} CD11b$^+$ monocytes populations in the cell cultures challenged the LPS changed (right panels) when compared with the PBS group

6. Examine the expression of CD86, CD38, and PD-L1 (Fig. 3) with FACS Canto II on exhausted monocytes generated by prolonged culture with high dose of LPS. Data can be analyzed with the FlowJo software.

4 Notes

1. For bone marrow harvest, work under the hood at all times and keep all media and reagents sterile.
2. The muscle and connective tissues attached to the bones should be removed as much as possible, because the presence of these tissues may reduce the viability of the bone marrow cells.
3. M-CSF powder should be dissolved in a complete RPMI-1640 medium to prepare the stock solution, and the stock solution should be aliquoted. One or several M-CSF aliquots can be used to prepare fresh M-CSF-containing medium before cell culture and each medium change.
4. Clean everything with 70% ethanol prior to using before bone marrow harvest.
5. During the bone marrow cell isolation, make sure to maintain the cells on the ice before finally culturing the cells in the incubator.

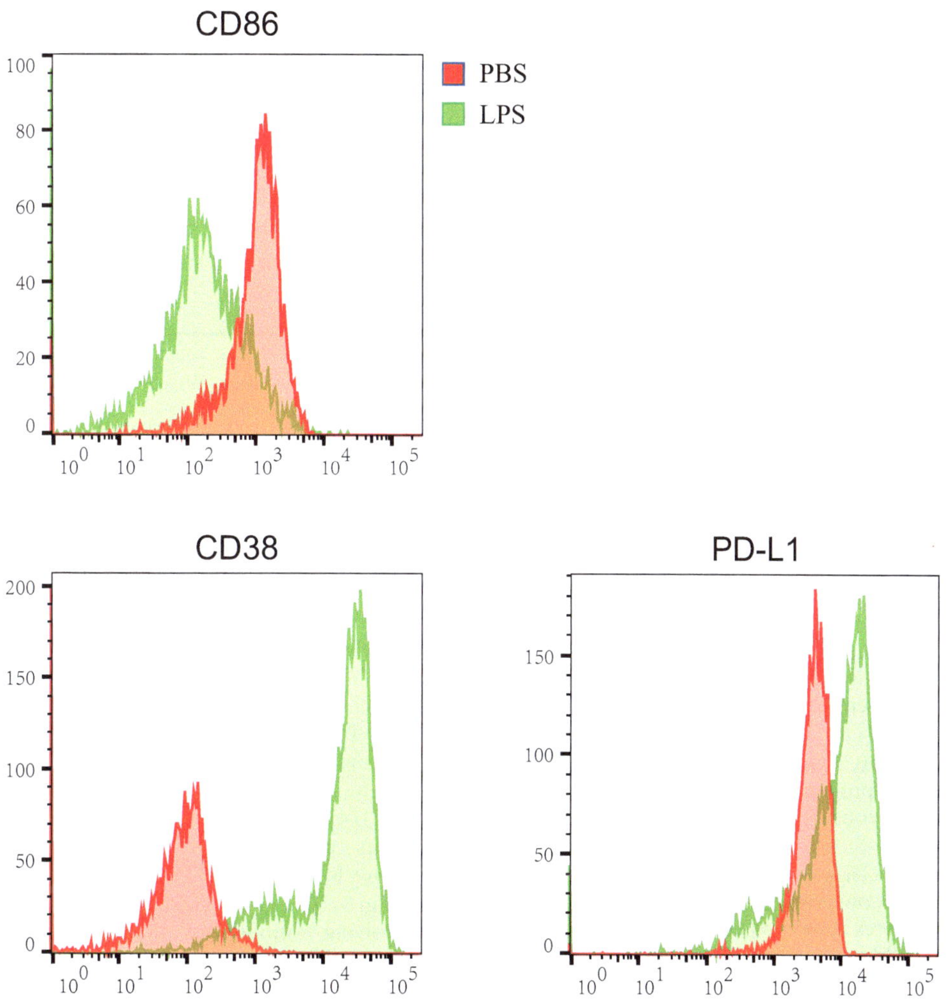

Fig. 3 Flow illustration of CD86, CD38, and PD-L1 by 5-day LPS stimulation. Bone marrow cells are cultured with M-CSF containing medium for 5 days. Cells are treated with either PBS or high-dose LPS (100 ng/mL) for 5 days in incubator. CD86 expression is reduced in the LPS group. Both CD38 and PD-L1 expression are upregulated in the LPS group

6. Pre-warm the cell culture medium (M-CSF-supplemented medium) to 37 °C before each medium change.
7. We use cell scraper instead of trypsin treatment to harvest monocytes at the end of cell culture, because trypsinization may cause damage to some cell surface molecules, which potentially affect the result of flow cytometry analysis. However, scraping monocytes may result in cell death. Therefore, it is crucial to incubate the cells in PBS on ice for 20 minutes before harvesting, and this can facilitate the detachment of monocytes.

8. Bone marrow cells can also be cultured in other containers (such as Petri dishes) if large numbers of memory monocytes are needed, and the volume of M-CSF-containing medium and LPS should be adjusted accordingly.
9. It should be noted that LPS from different *Escherichia coli* strains may have differential efficacies in priming memory monocytes. The strength of LPS should be carefully evaluated if using LPS from other *Escherichia coli* strains.

Acknowledgments

This study is supported by grants from the National Institutes of Health R01AI 432045 and R01 HL 412723 to LL.

References

1. Naler LB, Hsieh YP, Geng S, Zhou Z, Li L, Lu CZ (2022) Epigenomic and transcriptomic analyses reveal differences between low-grade inflammation and severe exhaustion in LPS-challenged murine monocytes. Commun Biol 5(102):1–17
2. Morris MC, Gilliam EA, Li L (2014) Innate immune programing by endotoxin and its pathological consequences. Front Immunol 5(680): 1–8
3. Yuan R, Geng S, Chen K, Diao N, Chu HW, Li L (2116) Low-grade inflammatory polarization of monocytes impairs wound healing. J Pathol 238: 571–583
4. Geng S, Chen C, Yuan R, Peng L, Maitra U, Diao N et al (2016) The persistence of low-grade inflammatory monocytes contributes to aggravated atherosclerosis. Nat Commun 7(13436):1–15
5. Pradhan K, Yi Z, Geng S, Li L (2021) Development of exhausted memory monocytes and underlying mechanisms. Front Immunol 12(778830):1–14
6. Cavaillon JM, Adrie C, Fitting C, Adib-Conquy M (2003) Endotoxin tolerance: is there a clinical relevance? J Endotoxin Res 9:101–107
7. Hutchins NA, Unsinger J, Hotchkiss RS, Ayala A (2014) The new normal: immunomodulatory agents against sepsis immune suppression. Trends Mol Med 20:224–233
8. Geng S, Zhang Y, Yi Z, Lu R, Li L (2021) Resolving monocytes generated through TRAM deletion attenuate atherosclerosis. JCI Insight 6(e149651):1–18
9. Lee J, Geng S, Li S, Li L (2021) Single cell RNA-Seq and machine learning reveal novel subpopulations in low-grade inflammatory monocytes with unique regulatory circuits. Front Immunol 12(627036):1–17

Chapter 6

A Robust In Vitro Co-culture Model for Studying the Intercellular Communication of Neutrophils

RuiCi Lin, Yao Zhang, Shuo Geng, and Liwu Li

Abstract

Communication among neutrophils plays critical roles during various phases of inflammatory responses, with clinical relevance to both acute and chronic inflammatory diseases. Despite its significance, underlying mechanisms are not well understood, due to the lack of an effective in vitro system to properly address this important question. Here we report a robust in vitro method to culture primary murine neutrophils derived from bone marrow, amenable for well-controlled studies of both neutrophil activation and intercellular communication among co-cultured neutrophils. This protocol can generate primary neutrophils with high purity and survival for an extended culture period, suitable for further phenotypic and functional analyses.

Key words Neutrophil, Polarization, Co-culture, Labeling, Intercellular communication

1 Introduction

Innate immune cells may propagate their memory states to neighboring cells through intercellular [1, 2]. The propagation of various innate immune phenotypes may be mediated through membrane contact and subsequent transfer of cytosolic materials including mitochondria and other subcellular organelles [3]. Neutrophil, the most abundant innate leukocyte and the first responder to inflammatory challenges, was recently shown to quickly adopt distinct functional phenotypes of resident leukocytes when transferred in vivo [4]. In order to gain a thorough understanding of neutrophil memory propagation, an efficient in vitro assay is warranted to enable an effective monitoring of intercellular communications among neighboring neutrophils.

Here, we describe an in vitro co-culture system providing a robust and easy-to-handle platform to examine inter-neutrophil communications. Donor neutrophils are purified from the murine

---

RuiCi Lin and Yao Zhang have equally contributed to this chapter.

Vijay Kumar (ed.), *Immune Homeostasis: Methods and Protocols*, Methods in Molecular Biology, vol. 2782, https://doi.org/10.1007/978-1-0716-3754-8_6,

bone marrow with percoll gradients and cultured in media supplemented with G-CSF which allows robust survival of cultured neutrophils within at least the 24-h culture period [1]. Cultured neutrophils can further be labeled with membrane-staining dyes, amenable to co-culture assays with recipient cells. This system allows flow cytometry–based evaluation to trace intercellular communication among donor and recipient neutrophils.

2 Materials

All cell culture solutions should be sterile and prepared in a class II biosafety cabinet. All reagents are stored at 4 °C, and waste materials are disposed of in compliance with the waste disposal regulations.

2.1 Bone Marrow Neutrophil Purification, Labeling, and Culture

1. 6-well cell culture plates, 50 mL conical tubes, 15 mL conical tubes, 70% ethanol, tweezers and scissors, 1 mL syringe (with 25-gauge needle), paper towels, 70 μm cell strainers, and 7 mL transfer pipettes.
2. Complete RPMI-1640 medium (Roswell Park Memorial Institute-1640 Medium): RPMI-1640 medium (Gibco™, # 11875085) supplemented with 10% FBS (Fetal Bovine Serum) (PEAK Serum, # PS-FB1), 1% penicillin-streptomycin (10,000 U/mL) (Gibco™, # 15140122), and 1% L-glutamine (200 mM) (Gibco™, # 25030081).
3. G-CSF (Granulocyte Colony Stimulating Factor) containing medium: Complete RPMI-1640 medium supplemented with 100 ng/mL G-CSF (PeproTech, # 250–05) and 10 mM HEPES (N-2-hydroxyethylpiperazine-N-2-ethane sulfonic acid) (Sigma-Aldrich, # 83264) (*see* **Note 1**).
4. Hanks' balanced salt solution (1× HBSS) (Sigma-Aldrich, # 6648), ACK (Ammonium-Chloride-Potassium) lysing buffer (Quality Biological, # 10128–806), phosphate-buffered saline (PBS) (Gibco™, # 10010023), HBSS (10×) (Gibco™, # 14185052).
5. 62.5% percoll gradient preparation: First, to make 100% isotonic percoll, mix 9 mL of percoll (Sigma-Aldrich, #P4937) with 1 mL 10×HBSS (Gibco™, #14185052). Then, further dilute 100% isotonic percoll with 1×HBSS (Sigma-Aldrich, #6648) to make 62.5% percoll: Mix 6.25 mL of the 100% percoll solution and 3.75 mL of 1× HBSS in a new 15 mL conical tube.
6. PKH26 red fluorescent cell linker mini kit for general cell membrane labeling (Sigma-Aldrich, # MINI26).

2.2 Flow Cytometry Analyses

1. Flow cytometry staining (FACS) buffer: 2% FBS in HBSS.
2. Fc blocker (purified rat anti-mouse CD16/CD32, BD Biosciences, # 533142), propidium iodide (PI) (1.0 mg/mL solution in PBS, Invitrogen, # P3566), and APC (Allophycocyanin)/Cyanine7 anti-mouse/human CD11b (Biolegend, # 101226).

3 Methods

3.1 Bone Marrow Collection and Neutrophil Purification

1. Euthanize mice following an approved IACUC (Institutional Animal Care and Usage Committee) procedure. Under sterile conditions, collect both left and right tibias and femurs from mice. Place tibias and femurs into a well of a sterile 6-well cell culture plate containing ice-cold RPMI-1640 completed medium (*see* **Note 2**).
2. Remove fleshes from the tibias and femurs with tweezers and scissors. Cut both ends of the tibias and femurs to expose the bone marrow (BM) content and transfer the bones to a new well containing 2 mL of ice-cold 1× HBSS.
3. Carefully flush out the bone marrow from the bones into the well containing 1× HBSS, using the 25-gauge syringe filled with 1× HBSS. Repeat this step until all the bones become white.
4. Collect bone marrow cells (in 1× HBSS) into a 15 mL conical tube and spin down the cells (390 × g, 7 min, 4 °C).
5. Remove the supernatant and resuspend the cells in 1 mL of ACK lysis buffer, mix well, and incubate on ice for 5 min to lyse red blood cells.
6. Add 9 mL of ice-cold PBS to stop the lysis reaction. Spin down the cells (390 × g, 7 min, 4 °C). Remove the supernatant and resuspend the cells in 5 mL of 1× HBSS.
7. Filter the cells in suspension with a 70 μm cell strainer into a new 15 mL conical tube to remove clotted cells/debris.
8. In a new 15 mL tube, add 5 ml of 62.5% percoll, and then carefully overlay 5 mL of cell suspension (from **step 9**) on the top of 62.5% percoll with a 7 mL transfer pipette without disturbing the interface between two different solutions (there should be a clear separation of two liquids after the loading).
9. Carefully place the gradient tube into the centrifuge and spin down the cells (1100 g, 30 min, 4 °C, brake off) (*see* **Note 3**).
10. Carefully remove the gradient tube from the centrifuge. Neutrophils are at the bottom of the gradient. The neutrophil pellet is loose and cloudy. Discard the cells at the interface

between HBSS and 62.5% percoll to make collection easier (*see* **Note 4**).

11. Collect neutrophils from the bottom (about 1 mL) and transfer them to a new 15 mL tube.
12. Wash the cells with 10 mL of 1× HBSS and spin down the cells (450 g, 7 min, 4 °C).
13. Remove the supernatant. Wash the cells again with ice-cold RPMI-1640 medium and spin down the cells (390 × g, 7 min, 4 °C).
14. The neutrophils are ready for culture or functional assays (*see* **Note 5**).

3.2 Neutrophil Cell Membrane Labeling

1. Prepare the staining dye: Add 1 μL of the PKH26 cell linker to 500 μL of diluent C and mix well (*see* **Note 6**).
2. Resuspend cells in 500 μL of diluent C and add 500 μL of the staining dye (prepared from the previous step) into the cell suspension. Mix well by gentle pipetting and incubate at room temperature for 1 min (*see* **Note 7**).
3. Add 1 mL of FBS to the cell-and-staining dye mix and incubate at room temperature for 1 min. Then spin down the stained cells (390 × g, 7 min).
4. Transfer cells to a new tube and wash the stained cells three times with 5 mL of RPMI-1640 complete medium and spin down the stained cells (390 × g, 7 min).
5. Resuspend stained cells in complete RPMI-1640 medium supplemented with G-CSF. According to the cell count, adjust the cell concentration to 2 million cells per mL of culture medium (*see* **Note 8**).

3.3 Cell Culture and Neutrophil Co-culture

1. Add 3 mL of cell suspension to each well (6×10^6 cells/well) in 6-well plates and place the plates in a cell culture incubator (37 °C with humidified 5% CO_2 atmosphere).
2. After 24 h of cell culture, collect the neutrophils into a 15 mL conical tube. Wash the well with 2 mL of pre-warmed RPMI-1640 medium twice to maximum yield and collect the medium into the same tube, followed by spinning down the cells (390 × g, 7 min, room temperature, RT) (*see* **Note 9**).
3. Remove the supernatant, wash the cells two more times with 4 mL of pre-warmed RPMI-1640 medium, and spin down the cells (390 × g, 7 min, RT).
4. Resuspend cells in the pre-warmed culture medium and adjust the cell concentration to 2 million cells per mL of culture medium. These cells are ready for co-culture and will be referred to as donor cells later.

5. Harvest and purify neutrophils from the murine bone marrow again (as Subheading 3.1 instructed). Resuspend cells in the pre-warmed culture medium and adjust the cell concentration to 1 million cells per mL of culture medium. These cells are ready for co-culture and will be referred to as recipient cells later.
6. Place 1.5 mL of recipient cell suspension to a well (1.5×10^6 cells/well) in a 6-well plate followed by adding 1.5 mL of donor cell suspension to the same well (3×10^6 cells/well). The ratio of cell numbers of donor and recipient cells is 2:1. (Extra donor and recipient cells should be cultured in separate wells to serve as controls.)
7. Co-culture cells for another 24 h followed by analyzing the cell phenotype with flow cytometry.

3.4 Analyses

1. Collect cells from the 6-well cell culture plate into an ice-cold 15 mL conical tube. Wash the well with 2 mL of ice-cold RPMI1640 medium once and collect RPMI1640 medium into the same tube. Spin down the cells ($390 \times g$, 7 min, 4 °C).
2. Remove the supernatant and resuspend the cells in the blocker buffer (1:100 diluted Fc blocker in FACS buffer and 25 μL per staining sample). Incubate on ice for 10 min (*see* **Note 10**).
3. Add 25 μL antibody cocktail (1:100 diluted APC/Cyanine7 anti-mouse/human CD11b antibody in FACS buffer) to the cells (the final dilution of CD11b antibody is 1:200). Incubate on ice for 20 min with an aluminum foil cover to avoid direct light.
4. Add 200 μL FACS buffer to wash the cells and spin down the cells ($450 \times g$, 5 min, 4 °C). Remove the supernatant and resuspend the cells in 200 μL of FACS buffer containing PI (1:4000 dilution).
5. Load the sample on FACS Canto II (BD Biosciences) and use FlowJo software (Ashland, OR) to analyze the flow data (Fig. 1).

4 Notes

1. Culture medium supplemented with G-CSF should be freshly prepared before cell culture.
2. The entire cell culture procedure should be operated under a biosafety cabinet, and keep all medium and reagents sterile.
3. The "brake off" setting at the neutrophil purification step is critical to avoid the disturbance of the gradients.

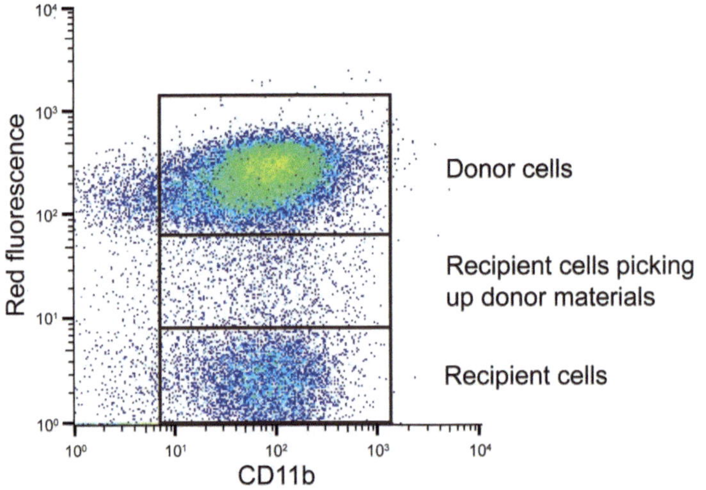

Fig. 1 Flow illustration of donor and recipient cells from the co-culture system Murine BM–derived neutrophils are labeled with red fluorescence and cultured in culture medium for 24 h (donor cells). Freshly harvested naïve murine BM neutrophils (recipient cells) are then co-cultured with donor cells for another 24 h. There are three separate cell populations with high CD11b expression: red-fluorescence-labeled cells (donor cells), intermediate cells (recipient cells which accepted membrane/cytosolic signals from donor cells), and non-labeled cells (recipient cells)

4. When removing the cells at the interface, leave at least 2 mL of the 62.5% percoll to avoid disturbing the neutrophil pellet.
5. Five to ten million neutrophils can be collected per mouse, and the purity is approximately 90% based on flow cytometry analysis with Ly6G and CD11b staining.
6. The labeling is conducted by following the manufacturer's instruction, and the amount of dye and the staining time are optimized for neutrophils. The amount of PKH26 cell linker dye stated in this protocol is enough to stain all the purified neutrophils from one C57BL/6 mouse's BM. If more cells are needed, the concentration of the dye needs to be further optimized.
7. Avoid leaving cells in diluent C for a long period of time.
8. In the presence of G-CSF, the neutrophils can be maintained well, and the percentage of live neutrophils after 24 h is above 95%, based on flow cytometry analysis with PI staining. Therefore, under this condition, the neutrophils can be cultured and used for testing the responses to certain stimuli in vitro).
9. Donor cells should be washed with the pre-warmed medium after the initial 24-hour incubation and resuspend in pre-warmed culture medium before co-culture in order to maintain cell function and reactivity.

10. For each staining sample (up to 1 million cells), 25 μL of blocker buffer (1:100 diluted Fc blocker in FACS buffer) is used and mixed with 25 μL of antibody cocktail (the final dilution of fluorescent conjugated antibody is 1:200 in general).

Acknowledgments

This study is supported by grants from the National Institutes of Health R01AI 432045 and R01 HL 412723 to LL.

References

1. Geng S, Zhang Y, Lee C, Li L (2019) Novel reprogramming of neutrophils modulates inflammation resolution during atherosclerosis. Sci Adv 5(2):eaav2309
2. Mittelbrunn M, Sanchez-Madrid F (2012) Intercellular communication: diverse structures for exchange of genetic information. Nat Rev Mol Cell Biol 13(5):328–335
3. van der Vlist M et al (2022) Macrophages transfer mitochondria to sensory neurons to resolve inflammatory pain. Neuron 110(4): 613–626 e619
4. Ballesteros I et al (2020) Co-option of neutrophil fates by tissue environments. Cell 183(5): 1282–1297 e1218

Chapter 7

Studying Monocyte-Myofibroblast Immunomodulation in a 3D Co-culture System

Avi Petroff

Abstract

Simple and reproducible 3D cell culture systems that mimic biological interactions within physiological tissues (biomimetics) can provide unique insight for scientific inquiries compared to 2D cell cultures. Fibroblast-populated collagen lattices (FPCLs) are commonly used for mimicking physiological collagen matrices, potentiating biomechanical stresses on embedded fibroblasts. Here, we describe a novel 3D co-culture model that incorporates human Tenon's capsule fibroblasts embedded in FPCLs co-cultured with THP-1 monocytes suspended in culture media. This method can be used for the assessment of cell-cell interactions in various stages of the wound healing process and can facilitate various types of immune cells in co-culture. This system can also be used to study pharmacological agents that may eventually improve clinical outcomes in patients affected by inflammatory disorders.

Key words Inflammation, Myofibroblasts, Monocytes, Glaucoma, Fibrosis, Co-culture, 3D Model

1 Introduction

The article titled "Understanding inflammation-associated ophthalmic pathologies: a novel 3D co-culture model of monocyte-myofibroblast immunomodulation" highlights a simple 3D co-culture system designed to develop an understanding of ocular myofibroblast interactions with circulating monocytes [1]. The purpose of this model was to facilitate increasingly physiologically relevant in vitro models of inflammation-associated ophthalmic pathologies.

The use of 3D co-culture systems such as the one described in this chapter can potentiate the discovery of novel molecules that could be feasible for clinical and therapeutic exploration. By observing changes in gene expression associated with inflammatory cytokine signaling between two cell types co-cultured in 3D, a better understanding of the progression of inflammation and inflammatory resolution can be elucidated.

Vijay Kumar (ed.), *Immune Homeostasis: Methods and Protocols*, Methods in Molecular Biology, vol. 2782, https://doi.org/10.1007/978-1-0716-3754-8_7,

The context of the model described in this chapter is derived from the original paper by Petroff et al. which assesses gene expression of human Tenon's capsule myofibroblasts and THP-1 monocytes in a 3D co-culture model that mimics the ocular environment post-glaucoma filtration surgery [1]. THP-1 monocytes are especially useful as an immortalized human monocytic cell line that is widely used to model inflammatory responses in vitro [1]. While this context directed much of the experimental design of this model, the parameters of the 3D co-culture model can be modified for myriad translational research objectives. The system can additionally be used to study pharmacological agents that may eventually improve clinical outcomes in patients affected by inflammatory disorders.

2 Materials

2.1 Preparation of Cell Media

Reagents required:

Dulbecco modified Eagle's minimal essential medium (DMEM, Gibco, Waltham, MA), sodium bicarbonate (Fisher Scientific), fetal bovine serum (FBS, Gibco, Waltham, MA), 2 mM L-glutamine (Gibco, Waltham, MA), 1% penicillin/streptomycin solution (Gibco, Waltham, MA), Roswell Park Memorial Institute (RPMI) 1640 media (ATCC, Manassas, VA), 0.05 mM β-mercaptoethanol (Thermo Fisher Scientific, Waltham, MA).

Preparation of cell culture media (alternative to commercial DMEM media).

1. According to the manufacturer instructions, weigh out 49.9 g of DMEM powder and dissolve in 4750 mL of milliQ water. Place the beaker on a magnetic stirrer and add a magnetic stir bar to mix.
2. Add 18.5 g of sodium bicarbonate (3.7 g/L).
3. The media needs to be pH of ~7.1 (because pH increases by 0.1–0.3 units after filtering the media. Optimal pH is 7.4).
4. Once the desired pH is obtained, aliquot 450 mL for subsequent steps.
5. Add 5 mL of L-glutamine.
6. Add 5 mL of antibiotic/antimycotic solution.

To make media containing 8% or 10% FBS:

7. Add 40 mL FBS to the media bottle to make DMEM media containing 8% FBS.
8. OR Add 50 mL FBS to the media bottle to make DMEM media containing 10% FBS.

To make FBS media containing 1% or 2% FBS:

9. Instead of **steps** 7 or **8**, add 5 mL FBS to the media bottle to make DMEM media containing 1% FBS.
10. Instead of **steps** 7 or **8**, add 10 mL FBS to the media bottle to make DMEM media containing 2% FBS.

To make RPMI-DMEM for THP-1 cells:

11. According to manufacturer instructions, add 0.05 mM β-mercaptoethanol to RPMI 1640-DMEM media (1:1 RPMI 1640:DMEM) supplemented with 2 mM L-glutamine (*see* **Note 1**).

2.2 Cell Culturing – Human Tenon's Capsule Fibroblasts (HTCFs)

1. DMEM media containing 8% FBS: Dulbecco modified Eagle's minimal essential medium (DMEM), sodium bicarbonate, fetal bovine serum (FBS), 2 mM L-glutamine, 1% penicillin/streptomycin solution, RPMI 1640 media, 0.05 mM β-mercaptoethanol.
2. Ice bucket.
3. Primary HTCF samples from liquid nitrogen storage.
4. 15 mL centrifuge tube
5. T75 cell culture flasks.

2.3 Cell Culturing THP-1 Cells

1. THP-1 cells from stock suspension.
2. 50 mL centrifuge tube
3. RPMI 1640-DMEM media (1:1 RPMI 1640:DMEM) containing 10% FBS: Dulbecco modified Eagle's minimal essential medium (DMEM), sodium bicarbonate, fetal bovine serum (FBS), 2 mM L-glutamine, 1% penicillin/streptomycin solution, RPMI 1640 media, 0.05 mM β-mercaptoethanol.
4. T150 cell culture flasks.

2.4 Cell Splitting – HTCFs

1. PBS solution.
2. Trypsin solution.
3. DMEM media containing 8% FBS: Dulbecco modified Eagle's minimal essential medium (DMEM), sodium bicarbonate, fetal bovine serum (FBS), 2 mM L-glutamine, 1% penicillin/streptomycin solution, RPMI 1640 media, 0.05 mM β-mercaptoethanol.
4. Suction device attached to a Pasteur pipette.
5. T75 cell culture flasks.

2.5 Cell Splitting – THP-1 Monocytes

1. PBS solution.
2. Trypsin solution.
3. RPMI 1640-DMEM (1:1 RPMI 1640:DMEM) containing 10% FBS: Dulbecco modified Eagle's minimal essential medium (DMEM), sodium bicarbonate, fetal bovine serum (FBS), 2 mM L-glutamine, 1% penicillin/streptomycin solution, RPMI 1640 media, 0.05 mM β-mercaptoethanol.
4. T150 cell culture flasks.
5. 50 mL centrifuge tube.

2.6 Cell Counting

1. DMEM media containing 8% FBS: Dulbecco modified Eagle's minimal essential medium (DMEM), sodium bicarbonate, fetal bovine serum (FBS), 2 mM L-glutamine, 1% penicillin/streptomycin solution, RPMI 1640 media, 0.05 mM β-mercaptoethanol.
2. RPMI 1640-DMEM (1:1 RPMI 1640:DMEM) containing 10% FBS: Dulbecco modified Eagle's minimal essential medium (DMEM), sodium bicarbonate, fetal bovine serum (FBS), 2 mM L-glutamine, 1% penicillin/streptomycin solution, RPMI 1640 media, 0.05 mM β-mercaptoethanol.
3. Trypsin solution.
4. 15 mL centrifuge tube
5. 50 mL centrifuge tube
6. Suction device attached to a Pasteur pipette.
7. Microcentrifuge tube.
8. PBS solution.
9. Trypan blue.
10. Hemocytometer.
11. Light microscope.

2.7 Preparation of Fibroblast-Populated Collagen Lattice (FPCL): 6-Well Plates

1. 6-well plates
2. 1.8 mg/mL sterile collagen
3. Waymouth media.
4. 0.34 M NaOH solution
5. DMEM media containing 8% FBS: Dulbecco modified Eagle's minimal essential medium (DMEM), sodium bicarbonate, fetal bovine serum (FBS), 2 mM L-glutamine, 1% penicillin/streptomycin solution, RPMI 1640 media, 0.05 mM β-mercaptoethanol.

6. DMEM media containing 2% FBS: Dulbecco modified Eagle's minimal essential medium (DMEM), sodium bicarbonate, fetal bovine serum (FBS), 2 mM L-glutamine, 1% penicillin/streptomycin solution, RPMI 1640 media, 0.05 mM β-mercaptoethanol.
7. PBS solution.
8. Trypsin solution.
9. Centrifuge tubes for each triplicate.
10. Inflammatory Cytomix (optional and customizable).

2.8 Incorporation of THP-1 Cells Overtop FPCL

1. FPCLs in 6-well plates.
2. Suction device attached to Pasteur pipette.
3. 3 μm-pore transwell inserts
4. RPMI-DMEM media (1:1 RPMI 1640:DMEM) containing 1% FBS: Dulbecco modified Eagle's minimal essential medium (DMEM), sodium bicarbonate, fetal bovine serum (FBS), 2 mM L-glutamine, 1% penicillin/streptomycin solution, RPMI 1640 media, 0.05 mM β-mercaptoethanol.

2.9 Sample Collection and RNA Extraction of Each Cell Type

1. Direct-zol RNA MiniPrep kit (Zymo Research).
2. 0.5 mg/mL collagenase I solution
3. 27.5G needles
4. Microcentrifuge tube.
5. NanoDrop spectrophotometer ND-1000.

2.10 RT-qPCR Analysis

1. High-Capacity Reverse cDNA Transcription kit (Applied Biosystems, Foster City, CA).
2. TaqMan Fast Advanced Master mix (Thermo Fisher Scientific, Waltham, MA).
3. QuantStudio 5 Real-Time qPCR System (Applied Biosystems, Foster City, CA).

3 Methods

Carry out all procedures at room temperature in a sterile environment (such as a biosafety cabinet) unless otherwise specified.

3.1 Cell Culturing – Primary Human Tenon's Capsule Fibroblasts (HTCFs) from Liquid Nitrogen Stock

1. Warm DMEM media containing 8% FBS in a 37 °C water bath for about 20 min and prepare an ice bucket prior to HTCF collection. Sterilize any container entering the biosafety cabinet (*see* **Note 2**).

2. Identify which primary HTCF sample will be used from liquid nitrogen storage and annotate this in any associated binder of stored cells.
3. While wearing protective eyewear and cryo-safe gloves, remove the desired cell vial and place into the ice bucket. Return any liquid nitrogen storage equipment to its original location.
4. Place the vial containing HTCFs into the water bath for 2–3 min (*see* **Note 3**).
5. Remove the vial from the water bath and gently wipe down the vial with a Kim wipe before bringing the vial into the biosafety cabinet (sterile environment).
6. Add the contents of the vial to a 15 mL tube with 5 mL of DMEM media containing 8% FBS and centrifuge at 1200 RPM for 5 min.
7. Remove the supernatant using a micropipette and discard the supernatant.
8. Resuspend the pelleted cells in 10 mL of fresh DMEM media containing 8% FBS.
9. Plate the resuspended HTCFs in T75 flasks and store in an incubator at 37 °C in a 5% CO_2 atmosphere.
10. HTCFs should be stored this way until experimental use or up to and including the fifth cell passage of splitting (*see* **Note 4**). Past the fifth cell passage, cells should be discarded as biohazardous waste.

3.2 Cell Culturing – THP-1 Monocytes

1. Obtain THP-1 monocytes from storage and use a pipette to move THP-1 cell suspension from the storage container into a 50 mL centrifuge tube.
2. Centrifuge at 1200 RPM for 5 min.
3. Remove the supernatant using a pipette for most of the volume and using a micropipette for as much residual supernatant as possible.
4. Resuspend pelleted cells in 50 mL of fresh RPMI-DMEM media containing 10% FBS.
5. Plate the resuspended THP-1 cells in T150 flasks and store in an incubator at 37 °C in a 5% CO_2 atmosphere.
6. THP-1 cells should be stored this way until experimental use, splitting cell cultures as needed.

3.3 Cell Splitting – HTCFs

1. Identify when media should be changed and/or cells should be split (*see* **Note 4**).
2. Warm PBS, trypsin, and DMEM media containing 8% FBS in a 37 °C water bath for 20 min.

3. Remove T75 flasks containing HTCFs from the incubator and aspirate media using a suction device attached to a Pasteur pipette (*see* **Note 5**).
4. Gently add 5–10 mL of PBS to each flask, tilt the flasks back and forth, and aspirate the PBS (*see* **Note 6**).
5. Add 2 mL of trypsin to each flask, tilting the flasks back and forth to ensure coverage of the entire bottom of the flask.
6. Incubate flasks at 37 °C for about 5 min (*see* **Note 7**).
7. Confirm that most HTCFs in each flask have detached from the T75 flask using a light microscope.
8. Add equal volumes (2 mL) of DMEM media containing 8% FBS to each flask to inactivate the trypsin.
9. Aliquot equal volumes of media containing suspended HTCFs across a desired number of new T75 flasks ("splitting").
10. Top up the volume of each new T75 flask to 10 mL with DMEM containing 8% FBS and place each new flask back into the 37 °C incubator (*see* **Note 8**).

3.4 Cell Splitting – THP-1 Cells

1. Identify when media should be changed and/or cells should be split (*see* **Note 4**).
2. Warm PBS, trypsin, and RPMI-DMEM containing 10% FBS in a 37 °C water bath for about 20 min.
3. Remove T150 flasks containing THP-1 cells from the incubator and transfer the suspended cells in media into a 50 mL centrifuge tube.
4. Centrifuge cell suspension at 1200 RPM for 5 min to pellet cells.
5. Aspirate supernatant media from the centrifuge tube using a suction device attached to a Pasteur pipette (*see* **Note 9**).
6. Resuspend pelleted cells in 2 mL of fresh RPMI-DMEM containing 10% FBS.
7. Aliquot equal volumes of media containing suspended THP-1 cells across a desired number of new T150 flasks ("splitting").
8. Top up the volume of each new T150 flask to 50 mL of fresh RPMI-DMEM containing 10% FBS and place each new flask back into the 37 °C incubator (*see* **Note 10**).

3.5 Cell Counting

1. For HTCFs, collect cells from T75 flasks via trypsinization (*see* Subheading 3.3) and inactivate trypsin by adding DMEM media containing 8% FBS in equal amounts to trypsin and transfer suspended HTCFs to a 15 mL centrifuge tube.

2. For THP-1 cells, collect cell suspension from T150 flasks and add the cell suspension to a 50 mL centrifuge tube.
3. *The following steps in this section apply to either cell type*: Centrifuge the cell suspension at 1200 RPM for 5 min to pellet cells.
4. Aspirate media using a suction device attached to a Pasteur pipette (*see* **Note 9**).
5. Resuspend cells in 1 mL of appropriate media (DMEM containing 8% FBS for HTCFs, RPMI-DMEM containing 10% FBS for THP-1 cells).
6. Use a micropipette to mix the cell suspension by moving the fluid up and down (*see* **Note 11**).
7. Remove 10 μL of cell suspension and add this volume to a microcentrifuge tube.
8. Add 20 μL of PBS and 10 μL of trypan blue to the microcentrifuge tube (*see* **Note 12**).
9. Clean a hemocytometer with 70% ethanol and load 10 μL of the diluted cell suspension onto this.
10. Using a standardized counting method, determine the average number of cells per box of the hemocytometer and calculate cell concentration (*see* **Note 13**).

3.6 Preparation of Fibroblast-Populated Collagen Lattice (FPCL): 6-Well Plates

1. Set up collagen matrix:

   Keep the following in an ice bucket: 1.8 mg/mL of sterile collagen and neutralization solution (see below). Neutralization solution: made up of 3 parts Waymouth media and 2 parts 0.34 M NaOH solution (*see* **Note 14**). Prepare quantities of each solution such that each of the six wells will contain 4 parts collagen and 1 part neutralization solution (*see* **Note 15**).
2. Warm up DMEM containing 8% FBS, DMEM containing 2% FBS, PBS, and trypsin in a 37 °C water bath for about 20 min prior to use.
3. Count HTCFs (*see* Subheading 3.5) and ensure that there are enough HTCFs for 2×10^5 HTCF cells per mL of collagen matrix per well.
4. Each well will contain 1000 μL of mixture of collagen and neutralizing solution (3000 μL per triplicate). Thus, for each triplicate, 2400 μL of collagen and 600 μL of neutralization solution will be used to maintain the 4:1 ratio.
5. Prepare centrifuges tubes initially filled with only collagen in bulk for each set of triplicates (e.g., two tubes containing 2400 μL of collagen per 6-well plate), ensuring all indicated wells have sufficient collagen ready to make the matrix.
6. Prepare the collagen matrix by mixing the 600 μL of neutralization solution to the 2400 μL of collagen solution for each

triplicate (three wells). Within the same centrifuge tube, add a volume of HTCF cell suspension from **step 3** such that there are 2×10^5 HTCF cells per mL of collagen matrix per well (*see* **Note 16**).

7. Mix quickly and thoroughly using up-and-down pipetting motions, and aliquot 1 mL of matrix solution containing HTCFs each into three wells of the 6-well plate. This yields one triplicate (three wells) of the FPCL.
8. Tilt the 6-well plates such that the collagen matrix containing HTCFs coats the bottom of the wells.
9. Once the 6-well plate is ready with two triplicates, incubate at 37 °C for 20 min to allow formation of the FPCL.
10. Add 1 mL DMEM containing 2% FBS to each well and incubate for 72 h at 37 °C (*see* **Note 17**).
11. For a subset of FPCLs, a Cytomix preparation can additionally be used (*see* **Note 16** and **18**).
12. After 72 h, the FPCL is prepared (*see* **Note 19**).

3.7 Incorporation of THP-1 Cells Overtop FPCL

1. After the 72-h preparation of the FPCLs, tilt each 6-well plate and aspirate media using a Pasteur pipette.
2. Prepare 3 μm-pore transwell inserts that will be placed atop the FPCL in each well of the 6-well plate (*see* **Note 20**).
3. Count THP-1 cells (*see* Subheading 3.5) and ensure that there are enough THP-1 cells for 0.5×10^5 THP-1 cells/mL in 4 mL of RPMI-DMEM containing 1% FBS.
4. Add 0.5×10^5 THP-1 cells/mL in 4 mL of RPMI-DMEM containing 1% FBS to each collagen matrix (*see* Table 1).
5. Incubate co-cultures for 72 h at 37 °C.

3.8 Sample Collection and RNA Extraction of Each Cell Type

1. Prepare the Direct-zol RNA MiniPrep kit (Zymo Research) to extract RNA from samples – use TRIzol reagent to lyse the samples. Follow the manufacturer's protocol.
2. Warm 0.5 mg/mL collagenase I to 37 °C in a water bath.
3. Lyse each sample type.

   THP-1 cells: Collect cell suspensions from within transwell inserts in each well, centrifuge each sample at 1200 RPM for 5 min to pellet cells, then resuspend each sample in 150–200 μL of TRIzol reagent to lyse cells according to the manufacturer's protocol. Homogenize each sample five times with a 27.5G needle (*see* **Note 21**).
4. HTCFs: Wash each well once with PBS. Using a pipette, collect the collagen matrix from each well, placing each matrix into microcentrifuge tubes (*see* **Note 22**). For each microcentrifuge tube:

Table 1
THP-1 and HTCF preparations used for RT-qPCR analysis

Sample number	Sample identification	Control used	Experimental conditions	Rationale for sample
1	THP-1 mono-culture ($-C_1$)	N/A	THP-1 cell mono-culture on collagen matrix containing no HTCFs	Negative control for gene expression comparison in THP-1 cells
2	THP-1 mono-culture + LPS ($+C_1$)	($-C_1$)	THP-1 cell mono-culture stimulated with 1 μg/mL LPS on collagen matrix containing no HTCFs	Positive control to simulate pathogen associated molecular pattern (PAMP) recognition during wound healing
3	THP-1 mono-culture + inflammatory Cytomix ($-C_2$)	($-C_1$)	THP-1 cell mono-culture on collagen matrix containing no HTCFs, but incubated with inflammatory Cytomix for 2 days prior to washout	Negative control to account for residual inflammatory Cytomix effects on THP-1 cells after FPCL washout
4	THP-1 co-culture	($-C_1$)	THP-1 cell co-culture on FPCL containing HTCFs	Compared to ($-C_1$) for assessment of co-culture effects on pro-inflammatory cytokine and chemokine gene expression
5	THP-1 co-culture + inflammatory Cytomix	($-C_1$)	THP-1 cell co-culture on FPCL containing HTCFs after inflammatory Cytomix incubation for 2 days	Compared to ($-C_1$) for assessment of co-culture effects on pro-inflammatory cytokine and chemokine gene expression in inflammatory conditions
6	HTCF mono-culture ($-C_3$)	N/A	HTCF mono-culture with 1% RPMI-DMEM containing no THP-1 cells for 3 days before washout and addition of fresh 1% RPMI-DMEM containing no THP-1 cells	Negative control for gene expression comparison in HTCFs
7	HTCF mono-culture + inflammatory Cytomix ($+C_2$)	($-C_3$)	HTCF mono-culture with 1% RPMI-DMEM containing 0.75 μg/mL inflammatory Cytomix for 2 days before washout and addition of fresh 1% RPMI-DMEM containing no THP-1 cells	Positive control to simulate activated myofibroblast phenotype during would healing after glaucoma filtration surgery
8	HTCF co-culture	($-C_3$)	HTCF mono-culture with 1% RPMI-DMEM containing no THP-1 cells for 3 days before washout and addition of fresh 1% RPMI-DMEM containing THP-1 cells for co-culture	Compared to ($-C_3$) for assessment of co-culture effects on pro-inflammatory cytokine and chemokine gene expression

(continued)

Table 1
(continued)

Sample number	Sample identification	Control used	Experimental conditions	Rationale for sample
9	HTCF co-culture + inflammatory Cytomix	($-C_3$)	HTCF mono-culture with 1% RPMI-DMEM containing 0.75 μg/mL inflammatory Cytomix for 2 days before washout and addition of fresh 1% RPMI-DMEM containing THP-1 cells for co-culture	Compared to ($-C_3$) for assessment of co-culture effects on pro-inflammatory cytokine and chemokine gene expression in inflammatory conditions

Samples were mono- or co-cultured for 72 h prior to collection. Samples were then used for RNA extraction and reverse transcription for qPCR quantification of gene expression in each cell type

5. Centrifuge at 2000 RPM for 10 min and discard supernatant.
6. Add 150 μL of pre-warmed 0.5 mg/mL collagenase I.
7. Incubate in a 37 °C water bath for 10 min, vortexing every 5 min.
8. Remove from water bath and immediately add 500 μL of TRIzol reagent.
9. Leave tubes at room temperature for 30–60 min to allow collagenase I to digest undissolved matrix.
10. Homogenize each tube five times with a 27.5G needle (*see* **Note 21**).
11. Centrifuge at 1200 RPM for 30 s to remove debris and transfer supernatant to a new microcentrifuge tube.
12. Using the collected THP-1 cell and HTCF samples in tubes, assess RNA concentration and purity using a NanoDrop spectrophotometer ND-1000, following the exact protocol from the Direct-zol RNA MiniPrep kit (Zymo Research, Irvine, CA).
13. Store extracted RNA samples at −80 °C until use for RT-qPCR analysis.

3.9 RT-qPCR Analysis

1. Reverse transcribe cDNA from total RNA using the High-Capacity Reverse cDNA Transcription kit (Applied Biosystems, Foster City, CA), in a Thermocycler programmed for 10 min at 25 °C, 120 min at 37 °C, and 5 min at 85 °C.
2. Use TaqMan Fast Advanced Master mix (Thermo Fisher Scientific, Waltham, MA) to perform qPCR in a QuantStudio 5 Real-Time qPCR System (Applied Biosystems, Foster City, CA) with 40 cycles of denaturation (95 °C for 1 min) and annealing/extension (60 °C for 20 s) (*see* **Note 23**).
3. Analyze data using desired parameters, ensuring proper templates, primers, and control (*see* **Note 24**).

4 Notes

1. More specific details on RPMI 1640 media can be found in the review paper describing THP-1 cells [2]. For this project, a 1:1 mix of RPMI 1640 media and DMEM was used for THP-1 cell culture.
2. For the entirety of this chapter, any container should be sterilized prior to entering the biosafety cabinet using 70% ethanol in a spray bottle, ensuring the container is completely sealed. This usually involves a few sprays on the surface of the closed container, moving it into the biosafety cabinet, placing it down, and waiting for the alcohol to evaporate completely before opening the container.
3. Ensure that the vial of HTCFs placed in the water bath does not reach 37 °C. This is done by only submerging the vial in the 37 °C water bath for only 2–3 min. Bear in mind that many cell samples in liquid nitrogen storage are often in dimethylsulfoxide, and so the cells will subsequently need to be centrifuged and collected, while leaving the supernatant. This will be outlined in the following steps.
4. After a few days in the incubator, cells will grow in their flasks until confluency and will need to be split into more plates to provide enough cells for co-culture experimentation. Media should be changed every 2–3 days (about twice a week), and cells should be "split" once they have reached about 80–90% confluency to avoid contact inhibition in cells.
5. When aspirating media with a Pasteur pipette, the best technique is to tilt the T75 flask such that media collects in an area that is convenient to aspirate with the Pasteur pipette without touching any HTCFs that are adherent to the bottom of the flask. This technique also maximizes the volume of media that can be aspirated. A similar technique can also be used for aspirating other fluids, such as PBS, during splitting of cells.
6. This "wash" of the HTCFs incorporates an arbitrary volume of PBS and can be adjusted to the experimenter's discretion. When adding PBS to the flask, try to aim the incoming PBS to an area of the flask that does not have any adherent HTCFs. This will prevent any HTCFs from being accidentally removed from the flask surface by the PBS wash.
7. The goal of this step is to remove all HTCFs that are adherent to the T75 flask, allowing the HTCFs to float in suspension. Try to strike a balance of giving the trypsin enough time to take effect, but also do not expose the HTCFs to trypsin for too long or they may die.

8. The "splitting" takes place in these final two steps, in which a small volume of media with high cell concentration is "seeded" onto new flasks, and then the flasks are topped up to 10 mL using appropriate culture media. This way, the cells have a new physical environment to grow without risk of confluency until the next time splitting is required.
9. When aspirating supernatant media to yield the pelleted cells, be extremely careful to aspirate only the volume of supernatant such that no cells are accidentally aspirated. The best technique to do this is to tilt the centrifuge tube and aspirate such that gravity collects the supernatant media just above the pelleted cells, ensuring aspiration of the media with minimal aspiration of cells.
10. T150 flasks containing THP-1 cells take relatively longer than T75 flasks containing HTCF fibroblasts to reach sufficient confluency requiring splitting. This is in part due to the larger flask allowing for more cell growth prior to confluency and also due to the THP-1 cells growing in suspension, contrasting HTCF growth adherent to the T75 flask surface. Keep this in mind as the splitting process is being conducted.
11. Mixing cells gently using a micropipette in this fashion allows for the most accurate counting of cells in a given volume of media. This is a crucial step that can cause erroneous calculations in subsequent cell culture steps if not performed. When in doubt, mix any solution used in this project prior to use.
12. This creates a 1:4 dilution of the cell suspension. If the cell pellet is noticeably large, you may need to dilute the cell suspension further (i.e., 1:8) for accurate cell counting.
13. Counting can be done in various ways. Our laboratory counted cells in the four corner boxes (4 × 4 grid) of the hemocytometer, and occasionally counted in the middle box (5 × 5 grid) in situations where there was a cell count of less than 100. After determining the total cell count of these boxes, divide by 4 (or 5 if you counted the middle box) to determine the average number of cells per box. Cell concentration then equals the average cell count per square x dilution factor/ 0.0001 (which is the volume of each square × the total volume of cell suspension of 1 mL).
14. The ratio of Waymouth media to NaOH in the neutralization solution could be changed to 2:2, but 3:2 was determined to be the optimal ratio. Also, ensure the pH of the matrix is approximately 7–7.2. Waymouth media can be made by dissolving a vial of Waymouth powder in 100 mL of distilled water and filtering the solution through a 0.2 μm syringe filter. The NaOH solution can be prepared by dissolving 0.68 g of NaOH pellets into 50 mL of distilled water and filtering through a 0.2 μm syringe filter.

15. Prepare 10% more volume than you expect to need to account for loss of volume, particularly with collagen. Always work in cold conditions, and check the pH of the matrix when mixing components (needs to be pink prior to addition into the 6-well plates).
16. The exact volume of HTCF cell suspension added to the matrix solution may vary, but the important factor is the concentration of HTCFs per mL of collagen matrix. It is recommended to achieve 2×10^5 HTCF cells per mL of collagen matrix per well. Additionally, this is the point when an "Inflammatory Cytomix" may be added (*see* **Note 18**).
17. DMEM containing 2% FBS is used in this step to allow for a more controlled rate of HTCF growth within the matrix. Alternative media conditions can be experimented with, but this method was found to be optimal within our laboratory.
18. To simplistically mimic an inflammatory microenvironment within the human eye, a subset of FPCLs were pretreated with an "Inflammatory Cytomix" (0.75 μg/mL each of TGFβ-1, TNF, IFNγ, and IL1β) for 24 h prior to co-culture with THP-1 monocytes [1]. This Cytomix can be modified to include any desired cytokine of interest. *See* Table 1.
19. The FPCL contains human Tenon's capsule fibroblasts. As these fibroblasts experience isometric biomechanical tension, they are purported to become activated into myofibroblasts (Petroff et al. [1]). It is strongly recommended that some cell-free collagen matrices are also prepared to serve as negative controls. This can be done by making the same collagen and neutralizing solution preparation, but omitting the HTCFs, allowing for only the collagen matrix polymerization without embedded HTCFs. More information on FPCLs can be found in the original published paper by Bell et al. [3].
20. HTCFs and THP-1 cells will be sharing the same co-culture media but will be unable to physically interact due to the transwell inserts. This ensures that the cells can communicate (i.e., via cytokine signaling), but the monocytes cannot penetrate the transwell inserts and enter the FPCL. It is recommended that FPCLs are visualized under a light microscope to confirm the maintenance of separation between the two cell types throughout the co-culture period. Our laboratory found that 3-μm-pore transwell inserts were optimal to allow for physical separation, but also allow for small molecule signaling between cell types.
21. Homogenizing cells in TRIzol through a 27.5G needle allows for thorough interaction between the reagent and the cells, maximizing the effect of the reagent in preparation for the

collection of cellular contents for analyses. Do this step slowly and carefully, especially since the TRIzol reagent should not be spilled.

22. This step seems strange, but it involves using the suction of the pipette to lift the globular collagen matrix containing HTCFs from the each well of the 6-well plate and transferring it into a microcentrifuge tube.
23. Samples should be tested in triplicate, and a non-template control should be included for each gene primer set master mix. To quantify gene expression, relative quantification (RQ) ΔΔCt should be used with the respective mono-culture controls ($-C_1$ and $-C_3$) as the calibrator/reference sample (*see* Table 1, Fig. 1).
24. The complete experimental workflow can be seen in Fig. 2. As indicated in the introduction, variations to the experimental design described in this chapter can allow for myriad experimental objectives to be explored. A similar 3D co-culture model is described by Gonga-Cavé et al. [4].

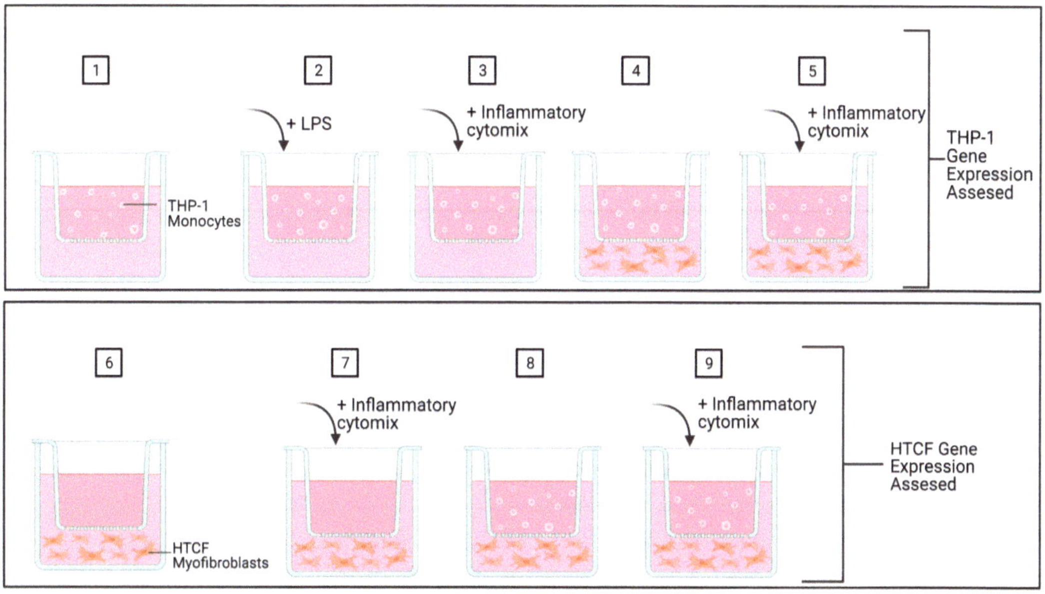

Fig. 1 THP-1 and HTCF samples used for RT-qPCR analysis. Samples shown in Table 1 are illustrated to highlight experimental conditions. (Image obtained from Petroff et al. [1])

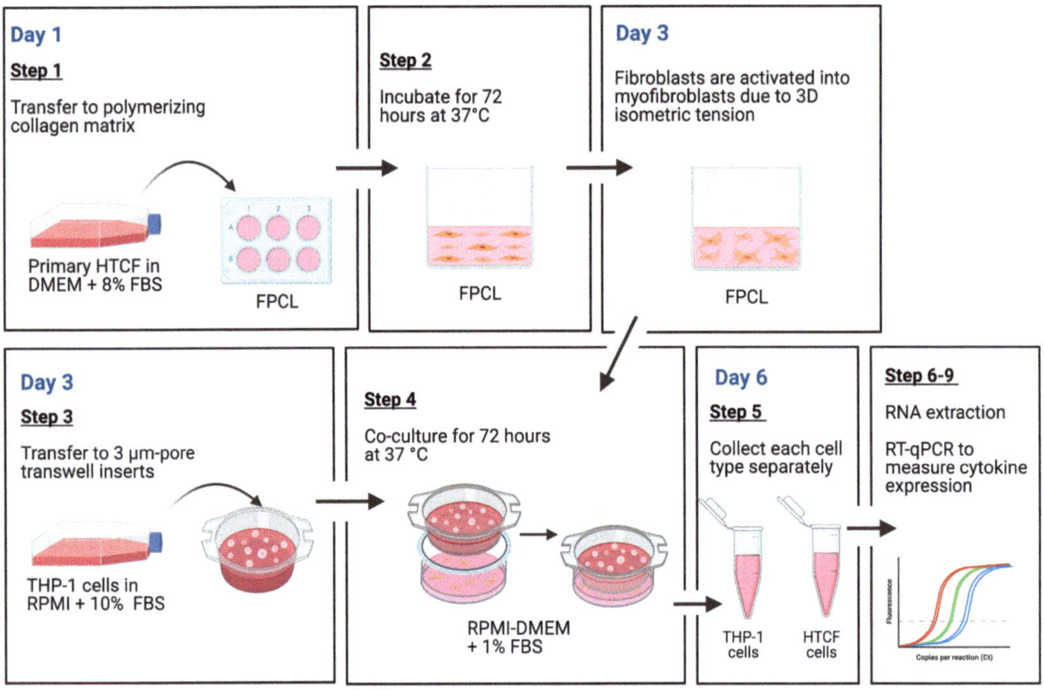

Fig. 2 Experimental workflow. Steps described within this chapter are illustrated to provide a simplistic overview of the experimental technique used for this project. Gene expression of inflammation-associated cytokines and chemokines can be assessed in THP-1 cells and human Tenon's capsule myofibroblasts (HTCFs) prior to RT-qPCR in the absence or presence of Inflammatory Cytomix, modified as desired. (Image obtained from Petroff et al. [1])

Acknowledgments

I acknowledge the students and staff of the O'Gorman and Hutnik laboratories for their support, communication, and mentorship for the duration of this project. I especially acknowledge Ana Pena Diaz, the laboratory technician of the O'Gorman research team, for collaboration on and refinement of experimental techniques described in this chapter.

References

1. Petroff A, Pena Diaz A, Armstrong JJ, Gonga-Cavé BC, Hutnik C, O'Gorman DB (2021) Understanding inflammation-associated ophthalmic pathologies: a novel 3D co-culture model of monocyte-myofibroblast immunomodulation. Ocul Immunol Inflamm:1–12
2. Chanput W, Mes JJ, Wichers HJ (2014) THP-1 cell line: an in vitro cell model for immune modulation approach. Int Immunopharmacol 23(1): 37–45
3. Bell E, Ivarsson B, Merrill C (1979) Production of a tissue-like structure by contraction of collagen lattices by human fibroblasts of different proliferative potential in vitro. Proc Natl Acad Sci U S A 76(3):1274–1278
4. Gonga-Cavé BC, Pena Diaz AM, O'Gorman DB (2021) Biomimetic analyses of interactions between macrophages and palmar fascia myofibroblasts derived from Dupuytren's disease reveal distinct inflammatory cytokine responses. Wound Repair Regen 29(4):627–636

Chapter 8

Endothelial Mitochondria-Associated Membranes (MAMs) Isolation by Percoll Step Gradients

Margaret Baldini, Cheng Zhang, and Jun Yu

Abstract

Mitochondria-associated membranes (MAMs) are regions where the endoplasmic reticulum (ER) interacts with mitochondria and regulate lipid trafficking, calcium signaling, ER stress, and inflammation activation. Isolation of MAMs from endothelial cells is vital for studying insight into the immune regulation of many inflammatory diseases. Endothelial cells (ECs) are critical innate immune cells due to their paracrine function of secreting interleukins, chemokines, cytokines, and growth factors, as well as expressing levels of pattern recognition receptors including toll-like receptors (TLRs). Furthermore, ECs regulate and recruit monocytes by expressing adhesion molecules including vascular cell adhesion molecule-1 (VCAM-1), intercellular adhesion molecule-1 (ICAM-1), P-selectin, and E-selectin, to facilitate monocyte diapedesis in areas of damage and inflammation. This protocol consists of step-by-step instructions on isolating pure MAMs and other subcellular fractions from endothelial cells, which is critical to understanding ER and mitochondria crosstalks in endothelial functions in health and disease.

Key words Mitochondria-associated membranes, Endothelial cells, Immunity, Inflammation, Mitochondria, Endoplasmic reticulum, Subcellular fractions

1 Introduction and Background

The integrity of the vascular wall is essential in maintaining proper vessel health and blood flow. In many inflammatory diseases such as atherosclerosis and hypertension, the endothelial layer of the vascular wall becomes activated and damaged, initiating a cascade of events that promote disease development [1]. Endothelial cells (ECs) compose a monolayer around the vascular lumen and regulate the vessel wall barrier permeability, cell proliferation, adhesion molecules, and inflammatory environments [1, 2]. Endothelial cell activation is an inflammatory response that heavily affects the vascular environment and elicits an immune response

Margaret Baldini, Cheng Zhang have equally contributed to this chapter.

Vijay Kumar (ed.), *Immune Homeostasis: Methods and Protocols*, Methods in Molecular Biology, vol. 2782,
https://doi.org/10.1007/978-1-0716-3754-8_8,

[2]. Furthermore, the concept that ECs can serve as immune cells in inflammatory diseases was coined over 80 years ago and has since been heavily researched [3, 4]. ECs function as critical innate immune cells by expressing pattern recognition receptors such as TLRs, as well as cytokines and chemokines, similar to macrophages [5, 6]. Appropriate EC function is crucial since most inflammatory, immune, and infectious diseases are initiated by EC dysfunction [7]. EC recognition of inflammatory stimuli, including danger-associated molecular patterns (DAMPs), will activate the expression of adhesion molecules (VCAM-1, ICAM-1, P-selectin, and E-selectin) and cytokines to facilitate monocyte diapedesis at areas of damage and inflammation [6–10].

Mitochondrial dysfunction has been directly linked to endothelial dysfunction and impaired vascular function through increased production of reactive oxygen species (ROS) [1]. Oxidative stress leads to endothelial wall injury, resulting in decreased endothelial nitric oxide (eNOS) activation and nitric oxide (NO) production and overproduction of ROS and proinflammatory cytokines [4, 11]. In these pathogenic states, mitochondria can facilitate the production of redox reactions that will enhance inflammation and endothelial dysfunction.

MAMs are specialized regions where mitochondria interact with the endoplasmic reticulum (ER) [12]. MAMs regulate lipid metabolism, ER stress, calcium signaling, protein transfer between the ER and mitochondria, and inflammasome formation [13–16]. This region was first described in 1958 and first visualized in the 1970s via electron microscopy. Since then, there have been numerous ways to study MAMs, including fluorophores and fluorescent proteins, electron tomography, and transmission electron microscopy [17–19]. However, the isolation of MAM fractions provides insight into the function of specific proteins that regulate lipid accumulation and trafficking [20–22]. Analysis of the subcellular fractions of cells and tissues can pinpoint the exact region of impaired function and is the only way to truly study the biological function of MAM regions [20–22]. For example, isolating subcellular fractions from a rat liver showed that reduced amounts of MAMs were directly linked to abnormal lipid export [20–22]. Because endothelial cells are critical inflammatory regulators, it is vital to understand the role of MAM fractions in the cellular response to oxidative stress.

This protocol provides guidance in isolating MAMs from human umbilical vein endothelial cells (HUVECs). This cell line is an in vitro model used to investigate endothelial cell functions, primarily related to but not limited to oxidative stress, inflammation, angiogenesis, and hypoxia [23, 24]. This protocol can be used for other types of ECs with minimal modification.

2 Materials

2.1 Reagents

- Fetal bovine serum (GE Life Sciences, #SH30396.03).
- HEPES (Sigma-Aldrich, #H3375).
- L-glutamine (100 mL) liquid (Gibco, #25030081).
- Penicillin–streptomycin (100 mL) (Gibco, #15140122).
- Percoll (Sigma-Aldrich, #P1644).
- Phosphatase inhibitor cocktail 2(100×) (Sigma-Aldrich, #P5726).
- Bovine serum albumin (BSA; Sigma-Aldrich, #A7906-50G).
- D-Mannitol (Sigma-Aldrich, #M9546-250G).
- Ethylenediaminetetraacetate dihydrate (EDTA; Sigma-Aldrich, #EDS-500G).
- Vascular cell basal medium (ATCC, # PCS-100-030).
- Endothelial Cell Growth kit-VEGF (ATCC, #PCS-100-041).
- Dulbecco's phosphate-buffered saline (D-PBS), liquid, without Ca^{2+} and Mg^{2+} (Thermo Fisher Scientific, # J67653.K2).
- Ethylene glycol-bis (2-amino-ethylether)-tetra-acetic acid (EGTA; Sigma-Aldrich, #E4378-25G).
- Protease inhibitor cocktail (25 tablets) (Roche, #11836170001).
- Sucrose (Sigma-Aldrich, #S7903-1KG).
- Trizma-Base (Sigma-Aldrich, #RDD008-2.5KG).
- Trypsin, 0.25% EDTA (Corning Life Sciences, #25-050-CI).

2.2 Equipment

- Cell culture dishes (150 × 20 mM) (Thermo Fisher Scientific, #150468).
- Tightly fitting glass Dounce homogenizer (Thermo Fisher Scientific, #1176F27).
- Loose fitting glass Dounce homogenizer (Thermo Fisher Scientific, #1176F27).
- Ultra-clear 14-mL polypropylene tubes (Beckman #331372) for SW40 rotor.
- Polycarbonate tubes with cap assembly (Beckman-ultra clear #344323) for 70-Ti rotor.
- Beckman Optima XE-100 ultracentrifuge.
- 70-Ti rotor (fixed angle, 8 × 39 mL, 70,000 r.p.m., 504,000 g) (Beckman #337922)
- SW 40 rotor (swinging bucket, 6 × 14 mL, 40,000 r.p.m., 285,000 g) (Beckman #331302).

- Eppendorf centrifuge 5810 R (15 amp version).
- FA-45-6-30 rotor (fixed angle, 6 × 50 mL, 14,200 r.p.m., 21,190 g) (Eppendorf #G439869N).
- A-4-62 rotor (swinging bucket, 4 × 250 mL, 4000 r.p.m., 3220 g) (Eppendorf #EPP-A-4-62).

3 Methods

3.1 Composition of Buffers

Note: Thoroughly wash each bottle and rinse with ddH_2O to avoid Ca^{2+} or Mg^{2+} contamination.

Buffer compositions have been separated into two tables, including buffers that can be prepared ahead of time (Table 1) and buffers that must be prepared fresh (Table 2).

3.2 Isolation of the Crude Mitochondrial Fraction (~2.5 h)

Recommended ten confluent plates (150 × 20 mM) for HUVEC are enough to isolate significant amounts of crude mitochondrial fraction needed to purify MAM and pure mitochondrial fractions.

Note: More cells are better.

1. Detach cells with trypsin and resuspend cells in the medium.
2. Wash cells with 30 mL PBS (with Ca^{2+} and Mg^{2+}) and centrifuge cells at 600 g for 5 min at 4 °C.
3. Wash cells one more time with 5 mL PBS (with Ca^{2+} and Mg^{2+}). Centrifuge cells at 600 g for 5 min at 4 °C.

Table 1
Composition of buffers that can be prepared ahead of time

Buffers needed (can be prepared ahead of time)	Buffer composition	Storage conditions
1-M Tris–HCl (pH 7.4)	Dissolve 121.14 g of Trizma base in 500 mL of bi-distilled water Adjust pH to 7.4 using HCl Bring the solution to 1 L with bi-distilled water	Store at 4 °C The solution remains stable for a long time (up to 6 months)
0.5-M HEPES (pH 7.4)	Dissolve 59.57 g of HEPES in 400 mL of bi-distilled water Adjust pH to 7.4 using KOH Bring the solution to 500 mL with bi-distilled water	Store at 4 °C The solution remains stable for a long time (up to 6 months)
100-mM EGTA (pH 7.4)	Dissolve 3.8 g of EGTA in 70 mL of bi-distilled water Adjust pH to 7.4 with KOH Bring the solution to 100 mL with bi-distilled water	Store at 4 °C The solution remains stable for a long time (up to 6 months)

These buffers can be stored for an extended period of time. After about 6 months, new buffers should be made

Table 2
Composition of buffers that must be prepared fresh

Buffers needed (must be prepared fresh)	Buffer composition	Storage conditions (onlys store for the duration of the experiment)
Starting buffer (SB)	225 mM mannitol, 75 mM sucrose, and 30 mM Tris–HCl pH 7.4 Dissolve 20.5 g of mannitol and 13 g of sucrose in 400 mL of bi-distilled water Add 15 mL of Tris–HCl (1 M pH 7.4) Leave the buffer for about 30 min at 4 °C to cool down Check the pH of the buffer and adjust with KOH (if too low) or HCl (if too high) Bring the solution to a final volume of 500 mL with bi-distilled water	Store at 4 °C
Isolation buffer for cells-1 (IBcells-1)	225 mM mannitol, 75 mM sucrose, 0.1 mM EGTA, and 30 mM Tris-HCl pH 7.4 Mix 150 mL of SB and 60 μl of 100 mM EGTA (pH 7.4).	Store at 4 °C
Isolation buffer for cells-2 (IBcells-2)	225 mM mannitol, 75 mM sucrose, and 30 mM Tris-HCl pH 7.4 IBcells-2 has the same composition as SB and is made up similarly	Store at 4 °C
MRB (mitochondria resuspending buffer)	250 mM mannitol, 5 mM HEPES (pH 7.4), and 0.5 mM EGTA To prepare 100 mL of MRB solution, dissolve 4.56 g of mannitol in 80 mL of bi-distilled water Add 1 mL of 0.5 M HEPES (pH 7.4) Leave the buffer for about 30 min at 4 °C to cool down Check the pH of the buffer and adjust with KOH (if too low) or HCl (if too high) Bring the solution to a final volume of 100 mL with bi-distilled water	Store at 4 °C
Percoll medium	225 mM mannitol, 25 mM HEPES (pH 7.4), 1 mM EGTA, and 30% percoll (vol/vol) Dissolve 2.052 g of mannitol in 25 ml of bi-distilled water Add 2.5 ml of 0.5 M HEPES (pH 7.4) and 0.5 mL of 100 mM EGTA (pH 7.4) Leave the buffer for about 30 min at 4 °C to cool down Check the pH of the buffer and adjust KOH (if too low) or HCl (if too high) Bring this basal solution to a final volume of 35 mL with bi-distilled water	Store at 4 °C

Fresh buffers are crucial to avoid Ca^{2+} contamination. Buffers can be kept at 4 for the duration of the experiment, but should be discarded once completed

4. Resuspend the cell pellet in 20 mL of ice-cold IBcells-1.
5. Homogenize cells using a tight Dounce homogenizer (pre-cold on ice). Check cell integrity under the microscope every 30 strokes (drop 10 μL to a slide and cover with a cover glass; use cells before homogenization as control). Finish homogenization when 80–90% of cell damage has been attained. (For HUVEC cells, 90–120 strokes is enough.)
6. Transfer the homogenate to a 50-mL polypropylene centrifugation tube and centrifuge at 600 g for 5 min at 4 °C. Collect the supernatant and discard the pellet (containing intact cells and nuclei).

   Note: This is to remove residual nuclei and non-broken cells.
7. Centrifuge the collected supernatant at 600 g for 5 min at 4 °C. Collect the supernatant and discard the pellet (if present).
8. Centrifuge the collected supernatant at 7000 g for 10 min at 4 °C. The pellet contains crude mitochondria. The supernatant should be kept as a cytosolic fraction containing lysosomes and microsomes.

   Pause point: For further separation of cytosolic and ER fractions, save the supernatant at 4 °C (or on ice) for up to 1.5 h. Centrifuge the supernatant at 20,000 g at 4 °C for 30 min. Further centrifugation of the supernatant at 100,000 g for 1 h. will result in the isolation of the ER (pellet) and cytosolic fraction (supernatant).
9. Gently resuspend the pellet containing mitochondria in 20 mL of ice-cold IBcells-2. Centrifuge at 7000 g for 10 min at 4 °C.
10. Discard the supernatant, resuspend the mitochondrial pellet in 20 mL of ice-cold IBcells-2, and centrifuge mitochondrial suspension at 10,000 g for 10 min at 4 °C.
11. Discard the supernatant and gently resuspend the crude mitochondrial pellet in 2 mL of ice-cold mitochondria resuspending buffer (MRB).

    Pause point: Save 0.2 mL of the crude mitochondrial fraction at −20 °C. It can be used for future western blot analysis of fractions.

3.3 Fractionation of Crude Mitochondria to Isolate the MAMs and Pure Mitochondria (~2.5 h)

1. Add 8 mL of percoll medium to the 14-mL thin-wall, polyallomer ultracentrifuge tubes (Beckman REF: 331372).
2. Layer suspension of mitochondria (in MRB buffer) on top of percoll medium. Afterward, gently layer the MRB solution (about 3.5 mL) on top of the mitochondrial suspension to fill up the centrifuge tube.

   Note: This step needs to be handled very gently.

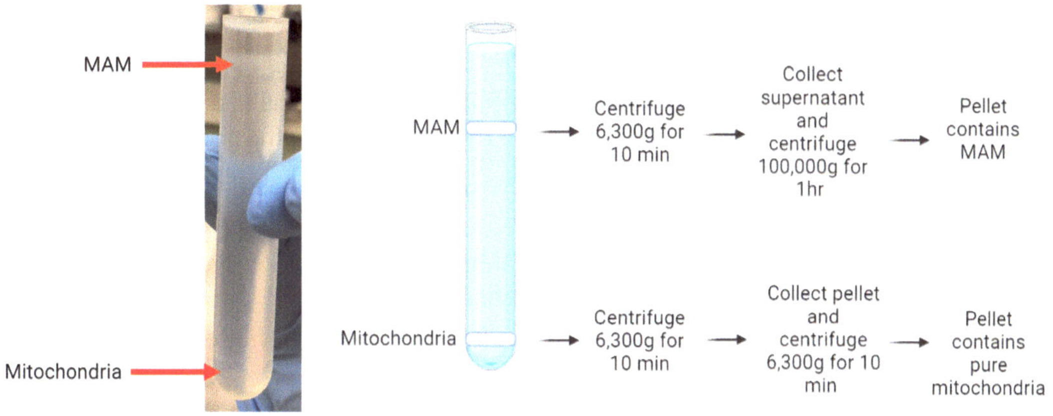

Fig. 1 MAM and pure mitochondria sub-fractionation from crude mitochondria. (The second portion of the figure was created using Biorender free)

3. Centrifuge at 95,000 g for 30 min at 4 °C. (Use the Type 40SW rotor, Beckman.) MAM is visible and localized toward the upper portion of the tube (diffused white band). A dense band containing purified mitochondria is localized approximately at the bottom of the ultracentrifuge tube (Fig. 1).
4. Collect the MAM fraction from the percoll gradient with a Pasteur pipette or by punching the tube wall with a syringe with a 22G needle. Dilute the MAM fraction ten times with MRB–MAM suspension.
5. Collect pure mitochondria band with a Pasteur pipette or by punching the tube wall with a 22G needle syringe. Dilute the mitochondria fraction ten times with MRB–mitochondria suspension.
6. Centrifuge MAM and mitochondria suspension at 6300 g for 10 min at 4 °C.
7. Transfer the MAM supernatant (discard the pellet) obtained in **step 6** to a polycarbonate tube (Beckman-ultra clear #344323) and centrifuge at 100,000 g for 1 h at 4 °C. (Use the Type 70Ti fixed rotor, Beckman.)
8. Discard the mitochondrial supernatant obtained in **step 6**, gently resuspend the pellet in 20 mL of MRB, and centrifuge at 6300 g for 10 min at 4 °C. Discard the supernatant and resuspend the pellet of pure mitochondria in a small volume of MRB (300–500 μL). This produces the pure mitochondria fraction. Store at −20 °C or immediately use for the western blotting.
9. Discard the supernatant (from **step 7**) and resuspend the MAM pellet in a small MRB volume (100–200 μL). This produces the MAM fraction. Store at −20 °C or immediately use for the western blotting.

4 Notes

Western blot analysis of markers for intracellular organelle-specific proteins will determine the success of the fractionation (Fig. 2).

- Calnexin is a MAM and ER marker and should be equally present.
- Mitofilin is a pure mitochondrion marker and should only be strongly present in that fraction.
- Facl4 is a MAM marker and should be highly enriched in this region.
- Tubulin is a cytosolic marker and should be present in the total lysate and cytosolic fractions. The absence of this marker in the pure and crude mitochondria, MAM, and ER indicates good fractionation.

All possible problems and tips on how to solve them have been synthesized into a table (Table 3).

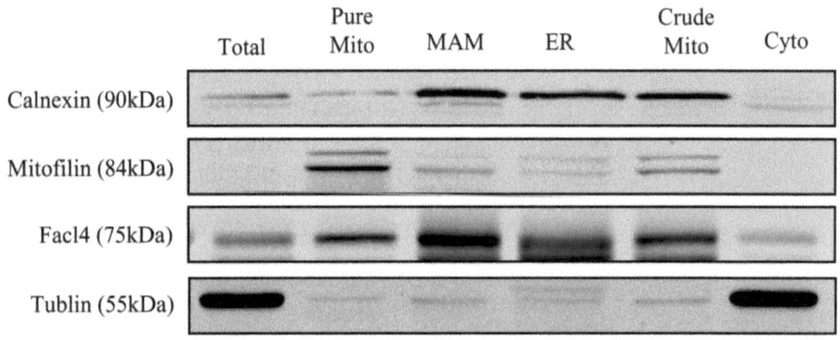

Fig. 2 Subcellular fraction separation and protein localization. Western blot comprised 12% sodium dodecyl sulfate-polyacrylamide gel electrophoresis (SDS-PAGE). A total of 30 μg of proteins were loaded on the gel, ran for 2 h at 100 V, and then transferred onto nitrocellulose membranes at 70 V for 4 h at 4 °C. Protein components of the subcellular fractions isolated from HUVECs are shown here. Total lysate, pure mitochondria, MAM, ER, crude mitochondria, and cytosolic. Proteins were detected with calnexin (MW: 90 kDa; 1:1000; ADI-SPA865); mitofilin (MW: 84 kDa; 1:2000; abcam #AB110329); Facl4 (MW: 75 kDa; 1:500; NOVUS #NBP2-16401); and tubulin (MW: 55 kDa; 1:200; Santa Cruz #SC-8035)

Table 3
Troubleshooting table describing common and possible problems, causes of problems, and solutions

Problem	Causes of problem	Solutions
Low amount of isolated MAM (little to no detection on western blot)	Not enough cells were used Less than 90% of cell damage was caused	Use more cells for isolation Homogenize longer (increase the force and number of strokes)
Low quality of MAM	MAM gradient band is broken and fragmented	Be very careful when layering the mitochondrial suspension of percoll medium Take off the break during the first centrifugation (**step 3** in "fractionation of crude mitochondria") Make sure to only take the MAM band and let no other material contaminate
Low amount or quality of isolated pure mitochondria	Not enough cells were used Less than 90% of cell damage was caused Contamination of pure mitochondria band	Use more cells for isolation Homogenize longer (increase the force and number of strokes) Make sure to only take the pure mitochondria band and let no other material contaminate

References

1. Yang YD et al (2019) Targeting mitochondria-associated membranes as a potential therapy against endothelial injury induced by hypoxia. J Cell Biochem 120:18967–18978. https://doi.org/10.1002/jcb.29220
2. Vanhoutte PM, Shimokawa H, Feletou M, Tang EH (2017) Endothelial dysfunction and vascular disease – a 30th anniversary update. Acta Physiol (Oxf) 219:22–96. https://doi.org/10.1111/apha.12646
3. Jungeblut CW, Berlot JA (1926) The rôle of the reticulo-endothelial system in immunity: II. The complement titer after blockade and the physiological regeneration of the reticulo-endothelial system as measured by reduction tests. J Exp Med 43(6):797–806. https://doi.org/10.1084/jem.43.6.797
4. Sturtzel C (2017) Endothelial cells. Adv Exp Med Biol 1003:71–91. https://doi.org/10.1007/978-3-319-57613-8_4
5. Drummer C et al (2021) Trained immunity and reactivity of macrophages and endothelial cells. Arterioscler Thromb Vasc Biol 41:1032–1046. https://doi.org/10.1161/ATVBAHA.120.315452
6. Shao Y et al (2020) Vascular endothelial cells and innate immunity. Arterioscler Thromb Vasc Biol 40:e138–e152. https://doi.org/10.1161/ATVBAHA.120.314330
7. Shao Y, Cheng Z, Li X et al (2014) Immunosuppressive/anti-inflammatory cytokines directly and indirectly inhibit endothelial dysfunction- a novel mechanism for maintaining vascular function. J Hematol Oncol. https://doi.org/10.1186/s13045-014-0080-6
8. Mai J, Virtue A, Shen J, Wang H, Yang XF (2013) An evolving new paradigm: endothelial cells–conditional innate immune cells. J Hematol Oncol 6:61. https://doi.org/10.1186/1756-8722-6-61
9. Joffre J, Hellman J (2021) Oxidative stress and endothelial dysfunction in sepsis and acute inflammation. Antioxid Redox Signal 35:1291–1307. https://doi.org/10.1089/ars.2021.0027
10. Reglero-Real N et al (2021) Autophagy modulates endothelial junctions to restrain neutrophil diapedesis during inflammation. Immunity 54:1989–2004.e1989. https://doi.org/10.1016/j.immuni.2021.07.012
11. Iba T, Levy JH (2018) Inflammation and thrombosis: roles of neutrophils, platelets and endothelial cells and their interactions in

thrombus formation during sepsis. J Thromb Haemost 16:231–241. https://doi.org/10.1111/jth.13911

12. Janikiewicz J et al (2018) Mitochondria-associated membranes in aging and senescence: structure, function, and dynamics. Cell Death Dis 9:332. https://doi.org/10.1038/s41419-017-0105-5
13. Perrone M et al (2020) The role of mitochondria-associated membranes in cellular homeostasis and diseases. Int Rev Cell Mol Biol 350:119–196. https://doi.org/10.1016/bs.ircmb.2019.11.002
14. Luan Y et al (2021) Structure and function of mitochondria-associated endoplasmic reticulum membranes (MAMs) and their role in cardiovascular diseases. Oxidative Med Cell Longev 2021:4578809. https://doi.org/10.1155/2021/4578809
15. Gao P, Yang W, Sun L (2020) Mitochondria-associated endoplasmic reticulum membranes (MAMs) and their prospective roles in kidney disease. Oxidative Med Cell Longev 2020:3120539. https://doi.org/10.1155/2020/3120539
16. Minamino T, Komuro I, Kitakaze M (2010) Endoplasmic reticulum stress as a therapeutic target in cardiovascular disease. Circ Res 107:1071–1082
17. De Copeland DA (1959) An association between mitochondria and the endoplasmic reticulum in cells of the pseudobranch gland of a teleost. J Biophys Biochem Cytol. https://doi.org/10.1083/jcb.5.3.393
18. Lopez-Crisosto C et al (2015) ER-to-mitochondria miscommunication and metabolic diseases. Biochim Biophys Acta 1852:2096–2105. https://doi.org/10.1016/j.bbadis.2015.07.011
19. Tubbs E, Rieusset J (2016) Study of endoplasmic reticulum and mitochondria interactions by in situ proximity ligation assay in fixed cells. J Vis Exp. https://doi.org/10.3791/54899
20. Wieckowski MR, Giorgi C, Lebiedzinska M, Duszynski J, Pinton P (2009) Isolation of mitochondria-associated membranes and mitochondria from animal tissues and cells. Nat Protoc 4:1582–1590. https://doi.org/10.1038/nprot.2009.151
21. Vance JE, Stone SJ, Faust JR (1997) Abnormalities in mitochondria-associated membranes and phospholipid biosynthetic enzymes in the mnd/mnd mouse model of neuronal ceroid lipofuscinosis. Biochim Biophys Acta 1344:1286–1299
22. Bassot A et al (2019) Regulation of mitochondria-associated membranes (MAMs) by NO/sGC/PKG participates in the control of hepatic insulin response. Cell 8. https://doi.org/10.3390/cells8111319
23. Gentile MT, Pastorino O, Bifulco M, Colucci-D'Amato L (2019) HUVEC tube-formation assay to evaluate the impact of natural products on angiogenesis. J Vis Exp 148. https://doi.org/10.3791/58591
24. Borkowska A, Ziolkowski W, Kaczor K, Herman-Antosiewicz A, Knap N, Wronska A, Antosiewicz J (2021) Homocysteine-induced decrease in HUVEC cells' resistance to oxidative stress is mediated by Akt-dependent changes in iron metabolism. Eur J Nutr:1619–1631. https://doi.org/10.1007/s00394-020-02360-8

Chapter 9

Analysis of Human Peripheral Blood Mononuclear Cells in Patients with Neurodegenerative Diseases

Mary Dover, Melika Madani, Milan Fiala, and Ketema Paul

Abstract

The role of immune system in the progression of neurodegenerative diseases has been studied for decades in animal models. However, invasive studies in human subjects remain controversial due to the heterogeneity of the presentation of different diagnostic categories at different stages of the disease. Peripheral blood mononuclear cells (PBMCs) contain immune cells including dendritic cells (DCs), monocytes, macrophages, and T lymphocytes. Isolating PBMCs from whole blood samples collected from patients provides a minimally invasive method for analyzing the immune system's function in patients with neurodegenerative diseases. By isolating single cell types from patients' peripheral blood, in vitro analyses can be conducted including RNA sequencing, immunofluorescence, and phagocytic analysis. In this chapter, we discuss PBMC separation and isolation of macrophages in pure culture in vitro. We also outline methods for performing RNA-seq on cultured macrophages and other techniques for investigating the role of macrophages in neurodegenerative disease pathophysiology.

Key words Peripheral blood mononuclear cells, Macrophages, RNA isolation, Phagocytosis, Neurodegenerative diseases, In vitro

1 Introduction

Peripheral blood is a good source of mononuclear cells [1] which are identified as blood cells with a round nucleus such as dendritic cells, monocytes, and macrophages [2]. These cells have immunological responses to various pathophysiological stimuli and therefore are used in immunologic studies, genomic analyses, and investigation of neurodegenerative disease progression [1, 3]. These analyses could contribute to characterizing the role of immune cells in diseases and infections and a better understanding of pathogenesis. Additionally, peripheral blood mononuclear cell (PBMC) examination aids in developing future therapies for various immune-related conditions [4].

Macrophages are a major cell population in the body and are a component of the mononuclear phagocyte system (MPS)

Vijay Kumar (ed.), *Immune Homeostasis: Methods and Protocols*, Methods in Molecular Biology, vol. 2782,
https://doi.org/10.1007/978-1-0716-3754-8_9,

[5]. These cells are critical for innate immunity, normal tissue development, and homeostasis [6] and can be studied through peripheral blood mononuclear cell (PBMC) isolation. Macrophages are derived from monocytes and are polarized into M1 and M2 types by cytokine signaling. Once differentiated, the M1-and M2-type macrophages play a role in combating infections and supporting homeostasis, respectively. While macrophages have beneficial anti-inflammatory responses against microbes and aggregated molecules, they can cause tissue damage in the lungs, heart, and brain when hyper-activated [7].

Neuroinflammation is a common pathological feature of many neurodegenerative diseases including Alzheimer's disease (AD) and amyotrophic lateral sclerosis (ALS) [8, 9]. Over the past few years, immunochemical demonstration of the central nervous system (CNS) invasion by peripheral blood macrophages in AD patients has resulted in support for the role of macrophages in neurodegenerative diseases. In AD, macrophages target amyloid-beta (Aβ) and P-tau, while in ALS, they target aggregated superoxide dismutase (SOD-1) and microbial nucleic acids [7]. Based on these functions, proper isolation of PBMCs including macrophages from the whole blood will allow for further investigation of the role of various cell types in neurodegenerative disease. Therefore, further understanding of the functions of these cells can be beneficial in therapy and understanding of neuroinflammation in neurodegenerative diseases such as AD and ALS [7].

For analysis of the immune system's function in response to neurodegenerative disease progression, a fundamental method is isolating PBMCs from whole blood samples. This isolation process is based on differences in density between PBMCs and other whole blood components. The underlying principles of this method of PBMC isolation have remained the same since the first separations of white blood cells using density gradient media in the 1960s. However, this technique has been refined in the following years [4] with increased efficiency and reproducibility. The procedure outlined in this chapter utilizes the Ficoll-Hypaque technique which isolates PBMCs from plasma and red blood cells (RBC) (Fig. 1). Ficoll-Hypaque is designed for mononuclear cell isolation and consists of Ficoll PM400 and sodium diatrizoate [10] which is denser than RBC but less dense than the desired PBMC buffy layer. Macrophages have been commonly isolated from mouse models rather than humans due to their ability to perform non-survival procedures and collect large amounts of tissue. However, processing human samples allows researchers to study how the immune system functions in patients and healthy controls directly, rather than through rodent models. To study macrophages in human samples, there are several limitations in place that ultimately make PBMC isolation the most efficient method. Most importantly, the Institutional Review Board (IRB) places limitations on

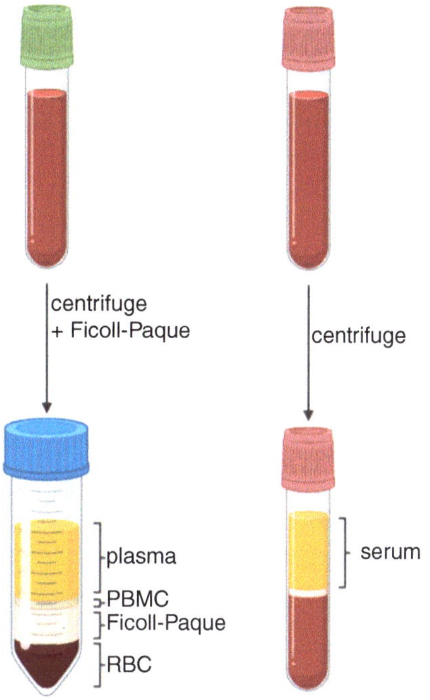

Fig. 1 A scheme depicting the density-gradient separation of anticoagulated (green cap) and coagulated (red cap) whole blood samples. (Image created with Biorender.com)

the degree of invasiveness of the procedure and typically does not allow for extensive tissue collection from volunteers. Thus, voluntary whole blood donation is an effective way to harvest many cells with minimal risk. Further isolation of single-cell types from these samples allows for the conduction of in vitro analyses such as RNA sequencing, immunofluorescence, and phagocytic analysis [11].

To evaluate the expression of genes in the inflammatory pathway, analysis of the PBMC and/or macrophage genome is an excellent tool. This chapter also outlines how to perform RNA isolation to be later analyzed by RNA-seq. The Qiagen RNeasy Mini Kit (Qiagen) will be utilized for this protocol due to its high yield and quality from repeated experiments; however, other RNA isolation kits may be used in its place.

Another mechanism for assessing PBMC function in patients with neurodegenerative diseases is via immunofluorescent staining. In Alzheimer's disease, Aβ is a cytotoxic protein implemented in disease progression. In vitro, the ability of macrophages to uptake Aβ can be simulated by introducing fluorescent Aβ to the culture and visualizing the level of uptake by the cells. This is considered a phagocytic analysis and can be performed relatively quickly. If desired, cells can be fixed at several time points over the day to analyze macrophage phagocytosis over a 24–48 h period. Live staining can also be utilized to measure cellular function in live or fixed cells. Cultured cells can be treated with a variety of fluorescent cell markers such as CellROX for reactive oxygen species, Mito-tracker for mitochondrial activity, etc. Staining for various activity markers such as mitochondrial activity can provide information about the overall function of a patient's macrophages alone or in the presence of a drug of interest.

Thus, this chapter outlines effective protocols for isolating PBMCs and single-cell types from patients' whole blood samples as a noninvasive method for analyzing the immune system's function in patients with neurodegenerative diseases.

2 Materials

2.1 Reagents

1. Ficoll-Hypaque PLUS (Sigma Aldrich). Store at 15–30 °C.
2. Phosphate-buffered saline (PBS) 1×: PBS 10×, deionized (DI) water. Prepare 1:10 PBS 10× in DI water. Store at 15–30 °C.
3. Bleach.
4. Macrophage medium: Iscove's modified Dulbecco's medium (IMDM) + GlutaMAX (Thermo Fisher), 1:1000 gentamicin. Store at 4 °C.
5. 0.4% trypan blue solution: trypan blue, DI water. Store at 15–30 °C.
6. 96–100% ethanol: ethanol, DI water.
7. 70% ethanol: ethanol, DI water.
8. Dimethyl sulfoxide (DMSO).
9. FAM-Aβ: 1 mg HiLyte fluor 488 amyloid-β_{1-42} (Anaspec, Fremont, CA), DMSO. Prepare 5:1 in DMSO for a final concentration of 200 μM, wrap in aluminum foil to shield from the light, and store at −20 °C.
10. CellROX Orange (Thermo Fisher). Store at −20 °C.
11. 2% Paraformaldehyde (PFA): 16% PFA, DI water. Dilute 125 mL 16% PFA in 875 mL DI water, and store in a cool, dry place.
12. Prolong Gold Antifade Mountant with DAPI (Thermo Fisher). Store at −5 to −30 °C and protect from light.

2.2 Supplies

1. Green-top anticoagulation glass phlebotomy tubes coated with sodium heparin, lithium heparin, and ammonium heparin.
2. Red-top glass phlebotomy tubes for serum collection.
3. 15 mL sterile polypropylene centrifugation tubes
4. 50 mL sterile polypropylene centrifugation tubes

5. 10 μL pipette and sterile fitted pipette tips
6. 200 μL pipettor and sterile fitted pipette tips
7. 1000 μL pipette and sterile fitted pipette tips
8. Hematocytometer.
9. 8-well glass cell culture slides with removable chamber
10. 6-well cell culture plates
11. Qiagen RNeasy Mini Kit containing buffer RNA lysis buffer (RLT), buffer RNA wash 1 (RW1), buffer RNA wash with ethanol (RPE), RNase-free water, Spin-away filters, 2 mL collection tubes, and 1.5 mL capped collection tubes.
12. Cell scrapers.
13. Aluminum foil.
14. Personal protective equipment (PPE) including gloves and lab coat.
15. 22 × 40 coverslips.

2.3 Equipment

1. Autopipette.
2. Centrifuge.
3. Microcentrifuge.
4. Mechanical balance scale.
5. Phase-contrast inverted microscope for cell culture viewing.
6. CO_2 cell culture incubator.
7. Deep freezer/dry ice.
8. Confocal microscope for cell imaging (or any desired immunofluorescent microscope).

3 Methods

All procedures should be performed under a fume hood at room temperature unless otherwise noted.

3.1 Isolation of PBMCs and Macrophages

This procedure utilizes Ficoll-Hypaque PLUS reagent which is a sterile medium that has been optimized for isolating human PBMC. All work with human cell lines should be completed under a certified fume hood to maintain a sterile environment.

1. From the human subject, collect 10 cc (10 mL) of whole blood in a green phlebotomy tube and 10 cc of whole blood in a red serum phlebotomy tube. Invert the green tube gently to ensure anticoagulation. Store the red-top tube containing blood at 4 ° C until ready for use.

2. In a 50 mL sterile centrifugation tube, dilute whole blood from the green tube 1:1 with PBS 1× for a total volume of ~20 mL. Mix by gentle inversion.
3. Add 4 mL each of Ficoll-Hypaque PLUS to two 15 mL sterile centrifugation tubes. Then, in a dropwise fashion, use an autopipette to carefully pipette 10 mL diluted whole blood mixture on top of the Ficoll-Hypaque layer in both 15 mL tubes (*see* **Note 1**). Once finished, carefully cap both 15 mL tubes and discard the empty 50 mL tube as biohazardous waste. Carefully transfer the 15 mL blood and Ficoll tubes to a mechanical balance scale to balance the tubes for centrifugation.
4. Centrifuge the balanced 15 mL tubes at 3000 rpm for 25 min with acceleration set to 9 and deceleration set to 0 (*see* **Note 2**).
5. Gently pipette out some of the topmost plasma layers from each 15 mL tube with an autopipette. This step will simply facilitate access to the thin buffy layer below. Be careful not to disturb the buffy layer below. Plasma can be discarded as biohazardous waste in bleach.
6. Isolate the cloudy buffy layer into a separate 15 mL tube using a 200 μL pipette (*see* **Note 3**). This layer contains the PBMC, so pipette gently to not damage the cells during transfer.
7. Dilute the PBMC with PBS 1× until the tube is filled to 14 mL. Gently cap the tube containing the diluted PBMC and again use the mechanical balance to balance out the centrifuge.
8. Centrifuge the PBMC at 1500 rpm for 7 min with 9 acceleration and 9 deceleration. After centrifugation, check for the presence of a PBMC pellet at the base of the tube. If no pellet is present, centrifuge again with the same settings. If a pellet is present, carefully discard the liquid from the tube in a 10% bleach solution, leaving behind roughly 1 mL PBS so as to not disturb the PBMC pellet.
9. Add 4 mL of warmed macrophage medium to the PBMC pellet and mix by inversion or flicking to break up the pellet (*see* **Note 4**). The total solution in the tube should now be roughly 5 mL; add more macrophage medium to add up to 5 mL if necessary.
10. *Cell counting*: Mix 10 μ PBMC with 90 μL trypan blue in a small sterile container. Add 10 μL of the mixture to a glass cell counting chamber and place a coverslip on top for viewing. Under a phase-contrast microscope, count the number of cells present within a 4 × 4 region of the cell counter. For more information on cell counting, please see **Note 5**.

    (# Cells within the 4 × 4 region) × (100,000) = # cells in 1 mL of the PBMC-macrophage medium mixture.

Table 1
PBMC plating[a]

Experiment type	Total volume media	Number of PBMCs/well
8-well plates	500uL	50,000 PBMCs/well
1-well plates	2 mL	1,000,000 PBMCs/well
6-well plates	2 mL	500,000 PBMCs/well
12-well plates	2 mL	125,000 PBMC/well
Glycoprotein	2 mL (use 15 mL tubes)	1,000,000 PBMCs/sample

[a]Plating of PBMCs is to be done in macrophage medium +12% autologous serum

Record the total number of cells in the 5 mL PBMC-macrophage medium mixture.

11. *Serum isolation*: Centrifuge the red-top phlebotomy tube containing coagulated blood at 1500 rpm for 5 min with 9 acceleration and 9 deceleration. Isolate the serum layer (Fig. 1) from the red-top tube into a sterile 15 mL tube. If there is a vague separation of the serum from the RBC layer following centrifugation, centrifuge again until the layers are distinct. Once the serum is isolated in a separate tube, dispose of the red-top tube and its contents as biohazardous waste.
12. *Plating cells* (Table 1) (*see* **Note 6**):
    (a) *8-well plates*: 8-well slides are used to plate cells that will eventually be treated for microscope imaging. Calculate how much PBMC is needed to plate 500,000 cells in each 8-well plate and dilute the PBMC with macrophage medium to the necessary concentration of cells. Supplement cells with 12% serum from **step 11**.
    (b) 6-well plates: 6-well plates are ideal when culturing cells for RNA isolation or other non-imaging procedures. Calculate how much PBMC is needed to plate 3 million cells per plate and dilute the PBMC with macrophage medium to the necessary concentration of cells. Supplement cells with 12% serum from **step 11**.
13. Incubate cells at 37 °C and 5% CO_2 in an incubator for 24 h or up to 2 weeks if isolating macrophages (*see* **Note 7**). Macrophages are adherent cells and are differentiated from monocytes. Macrophages that are incubated with adequate serum typically differentiate in 7–10 days. Non-adherent cells can be washed away from the well, leaving behind only differentiated macrophages for analysis.

3.2 RNA Isolation and Sequencing

If analyzing macrophages alone, cell cultures should be incubated for at least 7–10 days to allow macrophages to adhere and differentiate. At this point, all non-adherent cell types will be washed away with the medium.

RNA-seq analysis of the isolated RNA in this protocol has been processed by the University of California, Los Angeles CORE facility. See **Note 7** for the procedure utilized by the UCLA CORE lab in how to process isolated RNA samples for data analysis.

The following procedure has been adapted from the Qiagen RNA isolation kit manual with various modifications for improved sample collection.

1. Check that 6-well cell culture plates for differentiated macrophages/PBMC are present under a phase-contrast microscope (*see* **Note 8**).
2. Under a fume hood, gently aspirate and discard the medium from the desired cell culture wells (*see* **Note 9**). Do not leave cells dry for too long and proceed quickly to the next step. Cell culture medium should be disposed of as biohazardous waste.
3. Add 600 μL of the provided lysis buffer, buffer RLT, to each well of the culture plate and gently swirl to mix to lyse the cells. To disrupt any remaining adherent cells on the base of the well, use a cell scraper to gently pass over the bottom of the well.
4. On top of the buffer RLT in the wells, add 600 μL of 70% ethanol and mix with the buffer RLT by gently pipetting up and down several times.
5. Before transferring any sample to the collection tube, using a 1000 μL pipette, gently wash the solution over the base of the well while tilting the plate to ensure the complete removal of adherent cells. Then, using a 1000 μL pipette, transfer up to 700 μL of the mixed sample from each well into its own pink column placed in a 2 mL collection tube (*see* **Note 11**). Cap each pink column and centrifuge at 8000 × g for *15 s*. Discard all the flow-through in the 2 mL collection tube and, once empty, return it under the pink spin-away filter.
6. Using a 1000 μL pipette, add 700 μL of the stringent wash buffer, buffer RW1, to the same pink spin-away filter with the empty 2 mL collection tube beneath and centrifuge again at 8000 × g for *15 s*. Discard all flow-through and again return the empty collection tube under the pink spin-away filter.
7. Using a 1000 μL pipette, add 500 μL of the mild wash buffer, buffer RPE (*see* **Note 12**), to the same column with an empty 2 mL collection tube beneath and centrifuge at 8000 × g for *15 s*. Discard all flow-through and return the empty collection tube under the pink spin-away filter.

8. Again, add 500 μL of buffer RPE to the same column with an empty 2 mL collection tube beneath and centrifuge at 8000 × g this time for *2 min*.
9. Discard the 2 mL collection tube containing the flow-through and place the pink spin-away filter in a new 2 mL collection tube. Centrifuge again for *1 min* at 8000–10,000 × g to fully dry the column before proceeding.
10. Discard the 2 mL collection tube and place the pink column in an RNase-free 1.5 mL capped tube.
11. Using a 200 μL pipette, add 30 μL RNase-free water to the column and centrifuge for *1 min* at 8000 × g.
12. In the same column, repeat **step 11**, adding an additional 30 μL RNase-free water to the same column. Centrifuge again for 1 min at 8000 × g. The total volume in the collection tube should now equate to 60 μL and contain the isolated RNA.
13. Discard the pink column. Cap and label the 1.5 mL collection tube containing the isolated RNA and store at −80 °C until prepared for RNA-seq analysis.

3.3 Phagocytic Analysis

Cells cultured in 8-well glass slides must be used for this protocol to result in a fixed slide with cells to be visualized under a microscope. Additionally, if desired, supplements or treatments can be added to the wells at the same time as the Aβ and incubated to measure whether that specific additive has beneficial effects on phagocytosis. The following protocol outlines how to perform a phagocytosis analysis of PBMC or macrophages in culture.

1. Remove 8-well culture slides with cells from the incubator and check that cells are present or differentiated if observing macrophages.
2. Turn the lights off and perform the following steps in the dark to not overexpose the FAM-Aβ.
3. Using a 10 μL pipette, add 2 μL of FAM-Aβ to each well containing 500 μL of medium to a final concentration of 0.8 μM.
4. Cover with aluminum foil and incubate the 8-well plate for 24 h or for the length of time desired to measure phagocytosis.
5. After the desired length of time has passed (e.g., 24 h), ensure that the lights are off, and then gently aspirate and discard the medium from the wells. Do not leave cells dry for too long and proceed quickly to the next step. Cell culture medium should be disposed of as biohazardous waste.
6. Wash the wells once with 200 μL PBS 1×, swirl, and gently aspirate and remove (*see* **Note 13**).

7. Using a 200 μL pipette, add 150 μL 2% PFA to each well and incubate at room temperature for at least 20 min but no longer than 1 h. Optional: If desired, add 150 μL 0.2% PFA to each well and incubate at 4C for up to 1 week until prepared for the continuation of the protocol.
8. Gently aspirate and discard the PFA in a hazardous chemical waste container and follow institutional guidelines for adequate disposal.
9. Wash the wells once with 200 μL PBS 1×, swirl, and gently aspirate and remove.
10. Allow the wells to air dry for 5–10 min, and then carefully remove the plastic well chamber, leaving only the glass slide behind.
11. Add one drop of Prolong Gold Antifade Mountant with DAPI to each well square and place a 22 × 40 coverslip on top of the slide.
12. Store the slide flat in the dark at 4C for at least 24 h before imaging with confocal microscopy (*see* **Notes 14** and **15**).

3.4 Live Cell Staining

This protocol follows the same method of cellular preparation and imaging as the phagocytosis analysis, with some minor adjustments. Like the last protocol, cell cultures in an 8-well glass slide should be used for this protocol to facilitate microscope imaging. This specific protocol will utilize CellROX Orange (Thermo Fisher) as an example of a live cell stain to be used.

1. Remove 8-well culture slides with cells from the incubator and check that cells are present or differentiated if looking at macrophages.
2. Turn off the lights and perform **steps 3–6** in the dark to not overexpose the live cell stain.
3. Using a 10 μL pipette, add the appropriate volume of CellROX Orange to each well for a final concentration of 5 μM. Incubate for 30 min at 37 °C. If using a different live cell stain, be sure to follow company instructions on the final concentration and incubation time for each specific stain used.
4. Gently aspirate and remove all media from each well.
5. Wash once with PBS 1× (*see* **Note 13**).
6. Replace each well with 500 μL fresh macrophage medium and 12% autologous serum. Cover with aluminum foil to protect the cells from light and incubate for the desired length of time (e.g., 4 h) at 37 °C and 5% CO_2.
7. Gently aspirate and discard the medium from the wells. Do not leave cells dry for too long and proceed quickly to the next step. Cell culture medium should be disposed of as biohazardous waste.

8. Be careful not to touch the pipette tip to the bottom of the well to not disrupt the adherent cells.
9. Using a 200 μL pipette, add 150 μL 2% PFA to each well and incubate at room temperature for at least 20 min but no longer than 1 h.
10. Gently aspirate and discard the PFA in a hazardous chemical waste container and follow institutional guidelines for adequate disposal.
11. Wash the wells once with 200 μL PBS 1×, swirl, and gently aspirate and remove.
12. Allow the wells to air dry for 5–10 min, and then carefully remove the plastic well chamber, leaving only the glass slide behind.
13. Add one drop of Prolong Gold Antifade Mountant with DAPI to each well square and place a 22 × 40 coverslip on top of the slide.
14. Store the slide flat in the dark at 4C for at least 24 h before imaging with confocal microscopy (*see* **Note 15**).

4 Notes

1. When pipetting diluted whole blood on top of the Ficoll layer, use an autopipette that works slowly and smoothly. Initially, press down very slightly on the autopipette until the level of blood just starts to decrease and hold at this point. Place the tip of the pipette against the wall of the 15 mL centrifuge tube and hold the tube at a 45°angle to allow blood to gently flow down the side of the tube and lay on top of the Ficoll layer. Avoid pressing down suddenly or with any force on the autopipette as this may cause blood to shoot into the Ficoll layer. If that does occur, immediately discard the tube and prepare a fresh tube with 4 mL of Ficoll and reattempt. Any severe disruption to the Ficoll layer may impair the quality of the density gradient.
2. The Ficoll-Hypaque will aid in the creation of a density gradient with red blood cells (RBC) appearing at the base, the Ficoll-Hypaque solution above, then the buffy layer containing cells, and lastly the plasma layer on top. After centrifugation is complete, check tubes to ensure clear layer distinctions (Fig. 1). The buffy layer should appear as a thin cloudy band.
3. Upon the first few attempts of this step, pipetting out the thin, cloudy buffy layer may seem challenging. Visualize the pipette as a vacuum and check visually when pipetting to see the cloudy chunks in the layer getting sucked up into the pipette tip.

4. If the pellet is stubborn and will not un-adhere from the base of the centrifuge tube, brief spurts of vortexing the tube can be implemented. However, do not overuse the vortex as it can cause the cells to lyse.
5. Under a phase-contrast microscope, focus the lens so that a 4 × 4 region of the hematocytometer is within view. Be sure that you are looking at the region that also contains a faint 16 × 16 overlay. Within this region, count the number of cells, and count any cells on the top and right borders of the square, but not any on the bottom and left borders. This will give you a good estimate of the number of cells within your sample.
6. To avoid overcrowding, it is important not to plate the PBMC directly and instead to dilute the mixture with an additional macrophage medium. Additionally, supplementing cells with 12% autologous serum will promote cellular differentiation.
7. The following outlines the RNA-seq protocol utilized by the UCLA CORE facility for RNA processing. Assess total RNA quality on a TapeStation 2200 (Agilent Technologies) using an RNA HS ScreenTape assay. Measure the RNA concentration with Qubit RNA HS assay or Nanodrop device. Prepare libraries with the KAPA mRNA HyperPrep kit (Roche Sequencing) according to the manufacturer's recommendations on a custom Janus G3 Liquid Handler (Perkin-Elmer). Using 200 ng of total RNA as starting input for the library preparation capture poly-A RNA fragment to 200–300 nt. According to the manufacturer's protocol, perform reverse transcription, second-strand synthesis, and A-tailing. Use KAPA Dual Index Adapters (cat #KK8722) for the ligation and the reaction and purify twice using KAPA Pure beads. Perform 12 cycles of PCR amplification followed by purification according to the manufacturer's guidelines. Quantify final libraries using the Qubit dsDNA BR assay and average size with a D1000 ScreenTape assay on a 2200 TapeStation (Agilent Technologies). Dilute and sequence libraries on a NovaSeq 6000 (Illumina, Inc.) SP lane 2 × 50 bp. Process the RNA-sequencing data to remove adapter sequences and low-quality reads and then align to the human genome GCHr38 using STAR (v2.6+) [12]. Count data from HTSeq can be calculated [13] and normalized using DESeq2's median of ratios method [14] on any compatible coding software including Python, R, etc.
8. It has been found beneficial to check on the cells each day under a phase-contrast microscope and record the level of growth of the cells. If harvesting macrophages, check each day to assess the size of the cells. Initially, all cells in the plate will appear very tiny, but over the course of 1 week, the cells

will grow substantially, indicating that they are ready for analysis. To check that the cells have adhered, gently tap the plate while viewing them under a microscope, and see that the macrophages do not sway in the medium. Checking the cells frequently is also important to avoid any mold or fungus growth from spreading. If any contaminant is noticed early, seal off the infected well with parafilm and gently add bleach to the well. This will effectively contain the spread of the contaminant for at least 24 h.

9. Only aspirate the medium from the well that you would like to collect RNA from. For example, if you are performing an experiment that measures RNA expression over time, you will only need to aspirate the well that is dedicated to that time point. Another example is if you have treated your cells with a drug or chemical and are looking to measure the effects of the drug compared to an untreated sample. In this case, all of the wells you will be comparing will need to be isolated at the same time. Thus, decide how many and at what time you will be isolating RNA from the wells before beginning the RNA isolation experiment.
10. To increase collection, use a cell scraper and gently scrape any adherent cells off of the bottom of each well.
11. If performing RNA isolation on all wells, once the sample has been collected from every well in the pink column, the 6-well cell culture plate can be discarded as biohazardous waste. If only collecting RNA from some (not all) of the wells of the plate, after transferring the sample to the collection tube, return the 6-well plate to the incubator until prepared to isolate the rest of the RNA from the remaining wells.
12. If you are using the kit for the first time, add 96–100% ethanol 4:1 in the buffer RPE bottle provided. After this has been added, mix well by inversion.
13. Be careful not to touch the pipette tip to the bottom of the well to not disrupt the adherent cells.
14. The level of Aβ uptake by cells will be indicated by the level of fluorescent intensity of the cells when viewed under a fluorescent microscope. An example of images from a phagocytosis analysis can be visualized in Fig. 2.
15. Cells can be visualized via confocal microscopy or any fluorescent microscope. For higher-resolution imaging, confocal microscopy is suggested.

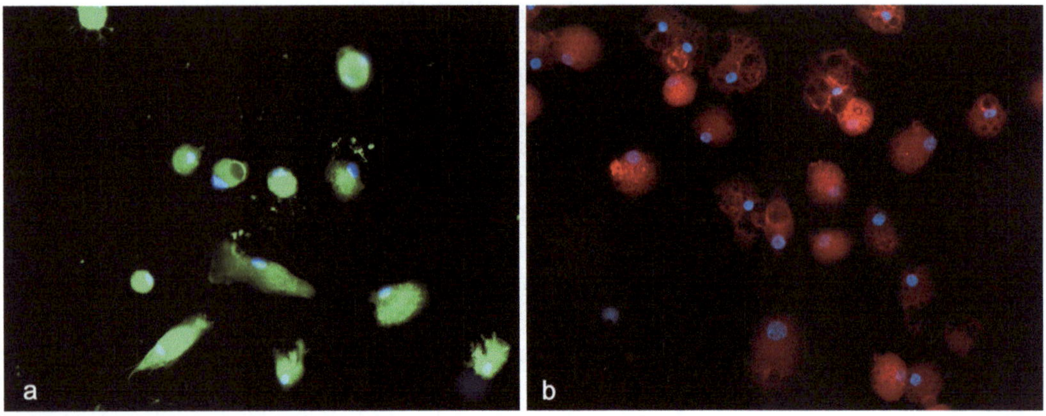

Fig. 2 Fluorescent phagocytic analysis and live cell stain of human isolated macrophages using protocol 3.2–3.3. (**a**) Phagocytic analysis displaying the level of uptake of FAM-Aβ (green) by macrophages: this image was obtained at 20× magnification and an exposure setting of 500 (blue: nuclei). (**b**) Live cell stain for reactive oxygen species (red) using CellROX Orange and DAPI for nuclei: this image was obtained at 20× magnification and an exposure setting of 200 (blue: nuclei)

References

1. Strober W (2001) Obtaining human peripheral blood cells. Curr Protoc Immunol Appendix 3: Appendix 3F. https://doi.org/10.1002/0471142735.ima03fs21
2. Pourahmad J, Salimi A (2015) Isolated human peripheral blood mononuclear cell (PBMC), a cost effective tool for predicting immunosuppressive effects of drugs and Xenobiotics. Iran J Pharm Res 14:979
3. Corkum CP, Ings DP, Burgess C et al (2015) Immune cell subsets and their gene expression profiles from human PBMC isolated by vacutainer cell preparation tube (CPT™) and standard density gradient. BMC Immunol 16:48. https://doi.org/10.1186/s12865-015-0113-0
4. Riedhammer C, Halbritter D, Weissert R (2016) Peripheral blood mononuclear cells: isolation, freezing, thawing, and culture. In: Weissert R (ed) Multiple sclerosis: methods and protocols. Springer, New York, pp 53–61
5. Hume DA (2006) The mononuclear phagocyte system. Curr Opin Immunol 18:49–53. https://doi.org/10.1016/j.coi.2005.11.008
6. Watanabe S, Alexander M, Misharin AV, Budinger GRS (2019) The role of macrophages in the resolution of inflammation. J Clin Invest 129:2619–2628. https://doi.org/10.1172/JCI124615
7. Dover M, Kishek M, Eddins M et al (2022) The role of M1- and M2-type macrophages in neurological and infectious diseases. IntechOpen
8. Krishnan D, Menon RN, Mathuranath PS, Gopala S (2020) A novel role for SHARPIN in amyloid-β phagocytosis and inflammation by peripheral blood-derived macrophages in Alzheimer's disease. Neurobiol Aging 93:131–141. https://doi.org/10.1016/j.neurobiolaging.2020.02.001
9. Beers DR, Appel SH (2019) Immune dysregulation in amyotrophic lateral sclerosis: mechanisms and emerging therapies. Lancet Neurol 18:211–220. https://doi.org/10.1016/S1474-4422(18)30394-6
10. Tan YS, Lei YL (2019) Isolation of tumor-infiltrating lymphocytes by Ficoll-Paque density gradient centrifugation. Methods Mol Biol 1960:93–99. https://doi.org/10.1007/978-1-4939-9167-9_8
11. Olivera-Perez HM, Lam L, Dang J et al (2017) Omega-3 fatty acids increase the unfolded protein response and improve amyloid-β phagocytosis by macrophages of patients with mild cognitive impairment. FASEB J 31:4359–4369. https://doi.org/10.1096/fj.201700290R
12. Dobin A, Davis CA, Schlesinger F et al (2013) STAR: ultrafast universal RNA-seq aligner. Bioinformatics 29:15–21. https://doi.org/10.1093/bioinformatics/bts635
13. Anders S, Pyl PT, Huber W (2015) HTSeq—a Python framework to work with high-throughput sequencing data. Bioinformatics 31:166–169. https://doi.org/10.1093/bioinformatics/btu638
14. Anders S, Huber W (2010) Differential expression analysis for sequence count data. Genome Biol 11:R106. https://doi.org/10.1186/gb-2010-11-10-r106

Chapter 10

Modulation and Determination of the Status of Inflammasomes in *Leishmania*-Infected Macrophages

Arunima Biswas, Anindita Bhattacharjee, and Pijush K. Das

Abstract

Leishmania, an intra-macrophage kinetoplastid parasite, modulates a vast array of defensive mechanisms of the host macrophages to create a comfortable environment for their survival. When the host encounters intracellular pathogens, a multimeric protein complex called NLRP3 inflammasome gets turned on, leading to caspase-1 activation-mediated maturation of IL-1β from its pro-form. However, *Leishmania* often manages to neutralize inflammasome activation by manipulating negative regulatory molecules of the host itself. Exhaustion of NLRP3 and pro-IL-1β result from decreased NF-κB activity in infection, which was attributed to increased expression of A20, a negative regulator of NF-κB signalling. Moreover, reactive oxygen species, another key requirement for inflammasome activation, are inhibited by mitochondrial uncoupling protein 2 (UCP2) which is upregulated by *Leishmania*. Inflammasome activation is a complex event and procedures involved in monitoring inflammasome activation need to be accurate and error-free. In this chapter, we summarize the protocol that includes various experimental procedures required for the determination of the status of inflammasomes in *Leishmania*-infected macrophages.

Key words *Leishmania*, Macrophage, Inflammasomes, NLRP3, ELISA, Immunoblot, Caspase-1, IL-1β

1 Introduction

One of the key mechanisms by which macrophages defend against any invading pathogen is by initiating the inflammatory response through the induction of chemokines and cytokines [1, 2] of which the pro-inflammatory cytokine IL-1β is the most important [3]. IL-1β, usually retained in the cytosol in an inactive form known as pro-IL-1β, gets proteolytically cleaved into IL-1β by caspase-1 [4, 5]. Caspase-1 is activated by the multi-protein inflammasome complex (NLRP3) comprising a sensor component which is usually a member of the nucleotide binding and oligomerization domain (NOD)-like receptor family (NLR) and an adaptor protein apoptosis-associated speck-like protein containing a CARD

Vijay Kumar (ed.), *Immune Homeostasis: Methods and Protocols*, Methods in Molecular Biology, vol. 2782, https://doi.org/10.1007/978-1-0716-3754-8_10,

(C-terminal caspase recruitment domain) or ASC (Adapter protein apoptosis associated speck-like protein containing a CARD) [6, 7].

To observe the status of macrophage inflammasome NLRP3 in *Leishmania* infection, the following changes/parameters are required to be assessed. Once activated, NLRP3 oligomerizes and promotes the polymerization of ASC via pyrin–pyrin interaction, exposing multiple caspase activation and recruitment domains, recruiting caspase-1 promoting its activity. The activation of the canonical NLRP3 inflammasome requires two signals. The first signal or priming is required to facilitate the increased expression of NLR protein and pro-IL-1β, as basal levels of these proteins are insufficient to trigger inflammasome activation. The second signal is required to induce the oligomerization and formation of inflammasome complex for pro-IL-1β processing. The priming signal depends on the recognition of pathogen-associated molecular patterns (PAMPs) through the pattern recognition receptors (PRRs) such as toll-like receptors (TLRs) leading to the transcription of pro-IL-1β and NLRP3 via NF-κB [8, 9]. TLR-induced NF-κB signalling is negatively regulated by the cytosolic inducible protein, A20 (also known as TNF-α–induced protein 3, TNFAIP3) [10], which has recently been shown by us to be upregulated by *Leishmania* [11].

The second activation signal which involves inflammasome dependent maturation of pro-IL-1β is derived from the generation of reactive oxygen species (ROS) [3]. ROS is primarily regulated by uncoupling protein 2 (UCP2), an inner mitochondrial membrane protein, which has recently been shown by us to be upregulated by *Leishmania* [12]. Suppression of inflammasome activation by *Leishmania* infection seems to be mediated by the upregulation of host A20 and UCP2 for inhibiting both the signals for inflammasome activation.

Therefore, to find out NLRP3 inflammasome activation, assessment of caspase 1, ROS and NF-κB is primarily essential. Moreover, since *Leishmania* downregulates NLRP3 inflammasome [13], a positive control is very much needed to assess inflammasome activation. Lipopolysachharide (LPS) and nigericin co-administered cells are used for this purpose. Usually, activation of various parameters is compared between LPS and nigericin-treated groups with LPS- and nigericin-treated *Leishmania*-infected groups. In this chapter, we discuss the experimental procedures carried out for analysing the mechanisms involved in the activation of inflammasome which is a complex and multistep process (Fig. 1). Moreover, since *Leishmania* usually inhibits inflammasome activation [13], to evaluate infection-mediated modulation, a positive control is necessary, and LPS plus nigericin was used as positive control to stimulate inflammasome activation. LPS plus nigericin treatment led to NF-κB mediated NLRP3 and pro-IL-1β transcription (signal 1 in Fig. 1) as well as ROS generation (second signal in Fig. 1), which facilitates the oligomerization

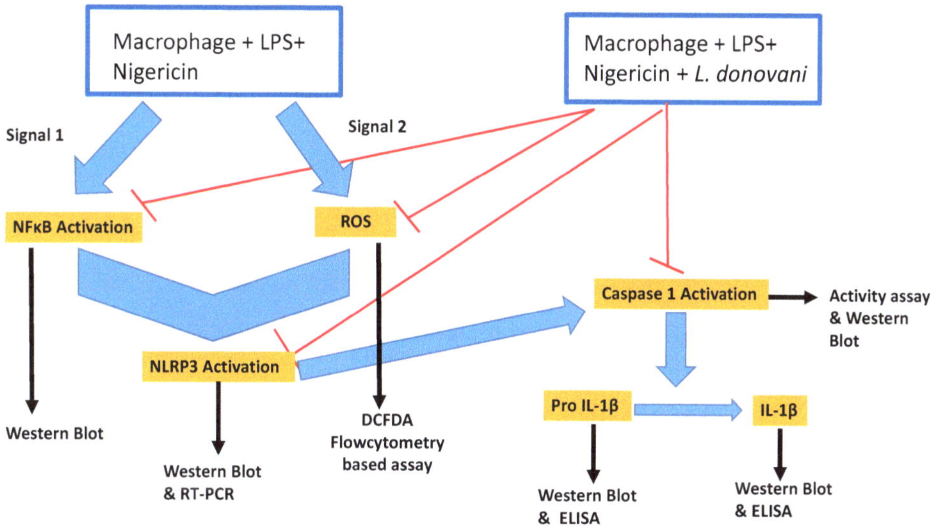

Fig. 1 When macrophages are treated with LPS plus nigericin, two signals are activated. Signal 1 induces NF-κB, which in turn activates NLRP3 transcription. Signal 2 leads to the production of ROS, which helps in the formation of an active inflammasome complex. The inflammasome complex then activates caspase 1, which further leads to the processing of mature IL-1β from its pro-form. On the contrary, *L. donovani* infected macrophages when subjected to LPS plus nigericin treatment, *Leishmania* inhibits signal 1 (by A20, not shown in the model) as well as signal 2 (by UCP2, not shown in the figure), ultimately hampering all the necessary steps of inflammasome activation and IL-1β production

and formation of an active inflammasome complex for the processing of pro-IL-1β to mature IL-1β by caspase 1. Although the level of IL-1β is the ultimate indicator of inflammasome activation, the process can be monitored at several steps, such as NF-κB activation, NLRP3 transcription, caspase 1 activity, ROS production and pro-IL-1β generation (Fig. 1).

2 Materials and Reagents Required to Observe the Status of Inflammasome Activation During Infection

2.1 Cell and Parasite Culture

1. *Leishmania donovani* parasites (MHOM/IN/1983/AG83).
2. M199 media (Thermo Fisher Scientific, Waltham, MA, USA).
3. L929-conditioned media (an alternative source of macrophage colony-stimulating factor, CSF-1).
4. RAW-264.7 cell line (adherent murine macrophage cell line, obtained from National Centre of Cell Science, Pune, India).
5. RPMI (Roswell Park Memorial Institute) 1640 media (Gibco, USA).
6. DMEM (Dulbecco's Modified Eagle Medium) low glucose (Gibco, USA).
7. Fetal bovine serum, heat inactivated, North American origin (Gibco, USA).

8. Penicillin and streptomycin (penicillin,10,000 U/mL/streptomycin, 100 µg/mL) (Gibco, USA).
9. Tissue culture flasks of various volumes and conical flasks.
10. Cell scrappers.

2.2 Experimental Animal Model and Primary Cell Cultures

1. Balb/c mice.
2. Bone marrow–derived macrophages (BMDM).
3. Peritoneal macrophages.
4. Splenic macrophages.
5. 1 µg/mL LPS.
6. Nigericin (20 µM).
7. Phosphate-buffered saline (PBS).
8. Giemsa Stain; for a volume of 500 mL: Giemsa powder, 3.8 g; absolute methanol, pure, high-grade, acetone-free, 250 mL; glycerol, high-grade, pure, 250 mL.
9. Dulbecco's phosphate-buffered saline (Sigma-Aldrich).

2.3 Reagents for RNA and ELISA

1. ELISA kit (R & D Systems).
2. RNeasy mini kit (Qiagen, Germany).
3. DNase 1 (Invitrogen, USA).
4. RT-PCR kit (Invitrogen, USA).

2.4 SDS PAGE and Western Blotting

1. Lysis buffer; 20 mM Tris, pH 7.4, 5 mM $MgCl_2$, 250 mM sucrose, protease inhibitor cocktail at 4 °C.
2. SDS.
3. 30% acrylamide.
4. SDS-PAGE running buffer (10×); 25 mM Tris base, 192 mM glycine, 0.1% SDS, pH 8.3.
5. SDS-PAGE separating buffer; 1.4 M Tris-HCl, pH 8.8.
6. SDS-PAGE stacking buffer; 4.95 M Tris-HCl, pH 6.8.
7. Ammonium per sulphate; 1 g APS in 10 mL of ddH_2O.
8. TEMED.
9. Sample loading buffer; 125 mM Tris-HCl, pH 6.8, 20% glycerol, 4% SDS, 10% ß-mercaptoethanol, 0.5 mg/mL bromophenol blue.
10. Pre-stained colour protein ladder (Thermo Fisher Scientific).
11. Tris-buffered saline (TBS), 10×; 1 M Tris-Cl, pH 7.5, NaCl.
12. Blocking solution; 5% BSA in 1× TBS-Tween.
13. Washing solution (1× TBS-Tween).
14. Proteins of interest (cytosolic extract).

15. Antibodies; diluted in the blocking solution, 1:1000 for primary antibodies and 1:10,000 for secondary antibodies. Antibodies from Santacruz, cell signalling and Sigma are generally used.
16. Substrates for chemiluminescent reactions (Pierce ECL western blotting substrate).

3 Materials

3.1 10% SDS (10 mL)

SDS – 1 g.

ddH$_2$0–10 mL.

3.2 RIPA Buffer (Lysis Buffer – 100 mL)

NaCl – 0.88 g.

Tris base – 0.60 g.

10% SDS – 1 mL

Triton X – 100 μL.

Make up the volume to 100 mL with ddH$_2$0 and check pH 7.4.

3.3 Loading Buffer (10 mL)

10% SDS – 1 mL

2-Mercaptoethanol – 1 mL

Glycerol – 2 mL.

Tris base – 0.15 g.

Bromophenol blue – 5–6 drops or more.

Add ddH$_2$0 to make up the volume.

3.4 Transfer Buffer (1 L)

Tris base – 5.81 g.

Glycine – 2.92 g.

Methanol – 200 mL.

SDS – 400 mg.

Add ddH$_2$0 to make up the volume.

3.5 TBS Buffer (10×, 1 L)

Tris base – 24.2 g.

NaCl – 80 g.

ddH$_2$0–800 mL,

Check pH 7.6 and make up the volume.

3.6 TBST Buffer (100 mL) (Wash Buffer)

10× TBS buffer – 10 mL

ddH$_2$0–90 mL,

Tween 20–100 mL.

3.7 Blocking Solution

BSA – 5 g.

TBST – 100 mL.

3.8 PBS (for 1 L)

Dissolve.

NaCl – 8 g.

KCl – 0.2 g.

Na_2HPO4–1.44 g.

KH_2PO4–0.24 g.

Distilled H_20–800 mL.

Adjust pH to 7.4.

Add H_2O to 1 L for volume make up.

Sterilize by autoclaving.

4 Methods

4.1 Parasite Culture

1. Administer *L. donovani* strain AG83 intravenously into the tail vein of the mice and maintain the Balb/c mice.
2. Sacrifice the mice after 6 weeks, obtain the amastigotes after dissecting the mice in a sterile condition and then macerating the spleen with the copper mesh. Assemble the contents and centrifuge them at 800 g to collect the splenic amastigotes and leave them for transformation in medium M199, supplemented with 10% FBS along with penicillin (50 U/mL) and streptomycin (50 μg/mL) at 22 °C in shaking condition for a period of 5–7 days. After the transformation to flagellate promastigotes, it is sub-cultured to obtain a *Leishmania* promastigote culture.

4.2 Isolation and Culture of BMDM

1. Sacrifice the mice (6–8 weeks old) and immerse them in 75% ethanol.
2. Clip and remove the skin from the lower part of the body.
3. Clean the remaining tissue from the femoral and pelvic bones thoroughly and separate the knee joint cautiously so as not to break the bones.
4. All the tissue associated with the bones should be properly removed to avoid any cross-contamination.
5. Immerse the bones in 75% ethanol for 5 min followed by their immersion in Dulbecco's phosphate-buffered saline (DPBS) for 5 min and then finally leave them in penicillin/streptomycin supplemented RPMI until the next step.
6. Cut off the end of each bone, then with the help of a 27-gauge needle/1 mL syringe, fill the bones with bone marrow growth

medium (RPMI 1640 supplemented with 100 U penicillin and 100 μg streptomycin/mL medium, 10% FBS, 5% horse serum and 10% L929-conditioned media) and expel the bone marrow from both ends of the bones with a jet of medium directed into a 15 mL cell culture dish.

7. The medium needs to be changed every 3 days, and whether all cells are attached to the plates is to be monitored frequently.
8. It usually takes about 10 days for all the cells to be attached to the dish, and then BMDM growth medium is used for continuation of culture.

4.3 Isolation and Culture of Murine Peritoneal Macrophages

1. For isolation of peritoneal macrophages, 3.8% Brewer's thioglycolate medium is to be prepared. Suspend 38 g of thioglycolate medium in 1000 mL of distilled water and boil the solution to dissolve the medium completely. Sterilize the solution by autoclaving at 121 °C for 15 min. This solution can be stored for up to 1 year in a dark and sterile condition.
2. Inject 1 mL of 3.8% Brewer's thioglycolate medium into the peritoneal cavity of each mouse using a 23-gauge needle/1 mL syringe.
3. After 3 days, each mouse is euthanized by cervical dislocation after proper anaesthesia by intraperitoneal injection of sodium pentobarbital (100 mg/kg body weight).
4. After washing the abdomen with 70% ethanol, perform a lateral incision along the bottom midline of the peritoneum. After pulling back the abdominal skin exposing the transparent peritoneal membrane, inject 5 mL of cold DPBS into the peritoneal cavity using a 20-gauge needle/5 mL syringe.
5. Gently massage the peritoneal cavity and then extract the fluid carefully without any damage to the organs with a syringe of 25-gauge needle. Remove the needle and collect the peritoneal fluid into 50 mL centrifuge tubes.
6. Centrifuge the fluid at 400 g for 10 min at 4 °C. Discard the supernatant and resuspend the pellet in RPMI 1640 medium with 10% FBS and penicillin-streptomycin.

4.4 Isolation and Culture of Splenocytes

1. Sacrifice the mice and sterilize the left side of each mouse by soaking them in 70% ethanol. Collect the spleen and macerate with a cell strainer or copper mesh in 1× PBS.
2. Transfer the cells into a centrifuge tube and centrifuge at 800 g for 3 min. Discard the supernatant and resuspend the pellet in DMEM and keep at room temperature for 5–10 min.
3. Discard the supernatant and resuspend the pellet in 3 mL of DMEM-10, discarding any dead cell mass.

4. Count the cells by haemocytometer and place them in culture overnight at 37 °C and 5% CO_2 and maintain the splenocytes in DMEM with 10% FBS.

4.5 Culture of RAW 264.7 Cells

Culture the adherent murine macrophage cell line RAW 264.7 in RPMI 1640 supplemented with 10% heat-inactivated FBS, 100 μg/mL streptomycin, and 100 U/mL penicillin and maintain the culture at 37 °C in a humidified incubator containing 5% CO_2.

4.6 Macrophage Infection with Leishmania

1. For in vitro experiments, infect the cultured macrophages with stationary phase *Leishmania* promastigotes at a ratio of ten parasites per macrophage.
2. After allowing the infection to proceed for 4 h, remove the non-phagocytosed parasites by washing the plates three times with PBS and allow the cells to grow for 20 h. After treating the cells with the required drugs and washing, fix the cells with methanol. Determine the number of intracellular parasites by staining with Giemsa stain and counting under a microscope.
3. For inflammasome activation, infect the cultured macrophages with *Leishmania* promastigotes for 4 h and simultaneously treat with 1 μg/mL LPS for 3 h 30 min followed by treatment with 20 μM nigericin for 30 min.
4. For in vivo experiments, inject female Balb/c mice (~20 g) with 10^7 stationary-phase promastigotes collected from cultures that had not been passaged more than three times via the tail vein. Quantify the infection by removing liver and spleen from infected mice at different times. Determine parasite burdens by Giemsa-stained impression smears. Express organ parasite burden as Leishman-Donovan units (LDU). One LDU is defined as the number of amastigotes/1000 nucleated cells x organ weight in grams. Carry out all the animal care and experimental procedures in strict accordance with the recommendations in the Guide for the Care and Use of Laboratory Animals of the National Institutes of Health.

4.7 Immunoblot Analysis

1. Lyse the macrophages with ice-cold lysis buffer/ RIPA buffer supplemented with protease cocktail inhibitor in a 1:100 ratio and incubate in an ice box for 40 min. Then centrifuge the mixture for 30 min at 13,000 rpm at 4 °C and collect and store the supernatant at −20 °C for further use.
2. Estimate and quantify the proteins using the Bradford reagent.
3. For performing the immunoblot assay, resolve an equal amount of protein by 10% SDS-PAGE by loading 30–40 μg of protein (mixed with loading buffer and heated for 5 min) and then transfer them to nitrocellulose membrane (Millipore-Sigma) by wet transfer method. After transfer using transfer

buffer, block the membrane with 5% BSA and dissolve in wash buffer (Tris-buffered saline (TBS) with 0.1% Tween 20) for 2 h at room temperature in shaking condition.

4. Incubate the membrane overnight with the appropriate primary antibody at the recommended dilution. Wash the membrane thoroughly with wash buffer and incubate with the corresponding alkaline phosphatase conjugated secondary antibody or horseradish peroxidase (HRP)-tagged secondary antibody for 2 h in the dark. For detection, use nitroblue tetrazolium (NBT)/5-bromo-4-chloro-3-indoyl phosphate (BCIP) tablets, according to the manufacturer's instructions.
5. For immunoblot analysis of IL-1β and caspase-1 p10, collect the cell supernatants prior to cell lysis at various time points and concentrate them with 72 μL trichloroacetic acid and 15 μL 10% cholic acid per ml of the supernatant.

4.8 Cytokine Analysis by ELISA

1. Perform sandwich ELISA to assess the levels of IL-1β in the single-cell suspension supernatant of spleen cells, peritoneal or BMDM or RAW 264.7 cells by using the sandwich ELISA kit (BD Biosciences, San Jose, CA, USA; R&D Systems, Minneapolis, MN, USA), as described by the manufacturers.

4.9 Real-Time PCR to Quantify the mRNA Levels of Pro-IL-1β and NLRP3

1. Isolate total RNA from splenocytes or BMDM/peritoneal macrophages or RAW 264.7 cells using the RNeasy minikit (Qiagen, Valencia, CA, USA) and with added DNase 1, according to the manufacturer's instructions.
2. Estimate the total amount of RNA by using a nanodrop spectrophotometer. Prepare cDNA from 1 μg of RNA with the SuperScript first-strand synthesis system for the RT-PCR kit (Thermo Fisher Scientific) as per the manufacturer's instructions.
3. Semiquantitative real-time PCR is to be performed on the ABI 7500 Fast real-time PCR system (Thermo Fisher Scientific) with the synthesized cDNA used as a template in the SYBR Green method. Use β-actin mRNA levels for normalization of each data, expression of which is measured as the fold change.

4.10 Caspase-1 Activity

Harvest the cells after infecting the RAW264.7 cells with *L. donovani* followed by appropriate treatments and wash twice with ice-cold 1 × PBS. Caspase-1 activity is then to be detected using caspase1 assay kit (Abcam, Cambridge, United Kingdom), as per the manufacturer's instructions.

4.11 Measurement of Reactive Oxygen Species (ROS)

1. Determine the intracellular ROS generation with an oxidant-sensitive green fluorescent dye, 2,7-dihydrodichlorofluoresceindiacetate (H_2DCFDA) (Thermo Fisher Scientific).
2. After *L. donovani* infection or appropriate treatment, harvest and resuspend the RAW 264.7 cells in H_2DCFDA (10 mM final concentration) and incubate for 15 min at 37 °C. Measure the fluorescence intensity of at least 10,000 cells using a FACS Canto II cell sorter (BD Biosciences) with a fluorescein isothiocyanate or FITC filter for determination of ROS generation, and the cells are to be gated on the basis of their fluorescent property.
3. The fluorescence intensity of H_2DCFDA can also be observed under a fluorescence microscope, and images may be captured.

References

1. Delbridge LM, O'Riordan MX (2007) Innate recognition of intracellular bacteria. Curr Opin Immunol 19:10–16
2. Medzhitov R (2010) Inflammation 2010: new adventures of an old flame. Cell 140:771–776
3. van de Veerdonk FL, Netea MG, Dinarello CA, Joosten LA (2011) Inflammasome activation and IL-1beta and IL-18 processing during infection. Trends Immunol 32:110–116
4. Eder C (2009) Mechanisms of interleukin-1beta release. Immunobiology 214:543–553
5. Li P, Allen H, Banerjee S, Franklin S, Herzog L et al (1995) Mice deficient in IL-1 beta-converting enzyme are defective in production of mature IL-1 beta and resistant to endotoxic shock. Cell 80:401–411
6. Martinon F, Mayor A, Tschopp J (2009) The inflammasomes: guardians of the body. Annu Rev Immunol 27:229–265
7. Bauernfeind F, Hornung V (2013) Of inflammasomes and pathogens--sensing of microbes by the inflammasome. EMBO Mol Med 5: 814–826
8. Latz E, Xiao TS, Stutz A (2013) Activation and regulation of the inflammasomes. Nat Rev Immunol 13:397–411
9. Bauernfeind FG, Horvath G, Stutz A, Alnemri ES, MacDonald K et al (2009) Cutting edge: NF-kappaB activating pattern recognition and cytokine receptors license NLRP3 inflammasome activation by regulating NLRP3 expression. J Immunol 183:787–791
10. Boone DL, Turer EE, Lee EG, Ahmad RC, Wheeler MT, Tsui C, Hurley P, Chien M, Chai S, Hitotsumatsu O, McNally E, Pickart C, Ma A (2004) The ubiquitin-modifying enzyme A20 is required for termination of toll-like receptor responses. Nat Immunol 5:1052–1060
11. Srivastav S, Kar S, Chande AG, Mukhopadhyaya R, Das PK (2012) *Leishmania donovani* exploits host deubiquitinating enzyme A20, a negative regulator of TLR signaling, to subvert host immune response. J Immunol 189:924–934
12. Basu Ball W, Kar S, Mukherjee M, Chande AG, Mukhopadhyaya R, Das PK (2011) Uncoupling protein 2 negatively regulates mitochondrial reactive oxygen species generation and induces phosphatase-mediated anti-inflammatory response in experimental visceral leishmaniasis. J Immunol 187:1322–1332
13. Gupta AK, Ghosh K, Palit S, Barua J, Das PK, Ukil A (2017) *Leishmania donovani* inhibits inflammasome-dependent macrophage activation by exploiting the negative regulatory proteins A20 and UCP2. FASEB J 31:5087–5101

Chapter 11

Fluid-Dynamic Culture of Tumour and Immune Cells for More Predictive Infiltration Studies and Immunotherapy Drug Screening

Maria Elisabetta Federica Palamà, Maurizio Aiello, and Silvia Scaglione

Abstract

Immunotherapies represent one of the current most promising challenges in cancer treatment. They are based on the boost of natural immune responses, aimed at cancer eradication. However, the success of immunotherapeutic approaches strictly depends on the interaction between immune cells and cancer cells. Preclinical drug tests currently available are poor in fully predicting the actual safety and efficacy of immunotherapeutic treatments under development. Indeed, conventional 2D cell culture underrepresents the complexity of the tumour microenvironment, while in vivo animal models lack in mimicking the human immune cell responses. In this context, predictability, reliability, and complete immune compatibility still represent challenges to overcome. For this aim, novel 3D, fully humanized in vitro cancer tissue models have been recently optimized by adopting emerging technologies, such as organ-on-chips (OOC) and 3D cancer cell-laden hydrogels. In particular, a novel multi-in vitro organ (MIVO) OOC platform has been recently adopted to culture 3D clinically relevant size cancer tissues under proper physiological culture conditions to investigate anti-cancer treatments and immune-tumour cell crosstalk.

The proposed immune-tumour OOC-based model offers a potential tool for accurately modelling human immune-related diseases and effectively assessing immunotherapy efficacy, finally offering promising experimental approaches for personalized medicine.

Key words Immune-organ-on-chip, 3D human tumour model, Immune cells, Cell extravasation, Cell infiltration

1 Introduction

Cancer-associated genomic instability determines the accumulation of numerous mutations, which can cause structural alterations and tumour antigen expression [1]. The immune system may recognize these mutated antigens as non-self and elicit a cellular immune response, playing a key role in immunosurveillance [2]. Cancer cells, on the other hand, have evolved several strategies to avoid immune vigilance, resulting in impaired effector function of

Vijay Kumar (ed.), *Immune Homeostasis: Methods and Protocols*, Methods in Molecular Biology, vol. 2782,
https://doi.org/10.1007/978-1-0716-3754-8_11,

immune cells and suppression of anti-tumour immune responses [3, 4].

A revolutionizing breakthrough in cancer treatment is represented by immunotherapies, aiming to boost the host's immune system towards cancer eradication [5]. The success of these strategies strictly depends on the interaction between immune cells and cancer cells. Innate immune cells (natural killer, eosinophils, basophils, and phagocytic cells) need to infiltrate into the complex tumour microenvironment (TME) and exert their tumour suppressor activity either by directly killing tumour cells or by triggering adaptive immune responses (promoted by lymphocytes, including B and T cells) [6, 7]. However, TME is a complex, heterogeneous, and dynamic entity, in which different cell types coexist and develop a mosaic of reciprocal interconnections, within an intricated three-dimensional vascularized matrix [8]. Due to the complexity of the tumour niche, the entire process might be greatly affected, leading to a negative outcome.

In this scenario, the development of reliable experimental models, allowing to understand the cellular and molecular underpinnings of immunotherapies is a crucial step to improve responsive rates and build up new efficient therapeutical strategies [9, 10].

To date, no tumour model can fully recapitulate the complexity of the TME. Traditional two-dimensional (2D) in vitro cell cultures still represent a gold standard in drug discovery screening because of the easy handling procedures and their cheap costs [11]. However, these too simplistic designs show low predictability and clinical translatability, largely due to the lack of tissue-level characteristics, such as vasculature or cell-cell and cell-matrix interactions [12].

On the other hand, in vivo animal models, such as immunodeficient mice xenografted with human cell lines or patient-derived tumour cells, may be adopted [13, 14]. However, although these models are useful for studying the systemic response to therapies, they still unsuccessfully predict the human immune responses [15]. In addition, in vivo xenografts do not accurately model the real behaviour of human tumours, due to the lack of a human stroma and its replacement with murine-derived tissues [10, 16].

A significant improvement in tumour modelling has come from the development of three-dimensional (3D) in vitro models, such as spheroids and organoids [17, 18]. Compared to conventional 2D tests, these models offer previously unattainable spatiotemporal control of microenvironmental elements that are crucial for tumour biology [19, 20]. Cancer cells exhibit a more in vivo–like behaviour in a 3D spatial architecture, such as upregulation of angiogenic factors due to the establishment of an oxygen gradient, a decreased proliferation, and a stronger chemoresistance [21, 22]. 3D microenvironment also plays a special role in cell-based immunotherapy testing, creating physical and cellular barriers to the infiltration of immune cells.

More recently, organ-on-chips are rapidly emerging as novel tools for controlled recreation of key cellular and molecular aspects, to improve predictability and translatability [23]. These innovative fluidic platforms have the potential to recapitulate the physiological blood flows and the fluidic connection among different organs, addressing the complexity and multi-dimensionality of human biology.

In this scenario, authors recently developed an organ-on-chip technology (MIVO®), an optically transparent cell culture chamber, compatible with the 24-well plate's size inserts, where clinically relevant size cancer tissue models are cultured under physiological flow conditions. The platform is suitable to investigate intra/ extravasation of circulating cells [24, 25] and to recapitulate a reliable and in vivo–like drug diffusion [26]. Moreover, MIVO® device provides an added value to immunotherapy testing, allowing to culture human immune cells directly in circulation, by imposing a physiological capillary blood flow, like in vivo. Cancer cells can be embedded in an extracellular natural tumour matrix and cultured into a permeable porous insert within the MIVO® chamber, to physically separate the tumour niche by the immune cells flow, like the in vivo physiological environment.

2 Materials

2.1 Reagents

Prepare all the required solutions as described below and, if required, sterilize them by using a 0.2 μm syringe filter, under a biological cabinet. Use the 3D-flowGel kit from React4life to prepare the three-dimensional cancer model.

1. Alginate-based solution: Dissolve the alginate powder from 3D-flowGel kit in buffer solution (physiological solution). Let the alginate powder dissolve under stirring at room temperature overnight. Filter and store at 4 °C.
2. Calcium-based cross-linking solution: Sterile and ready-to-use from 3D-flowGel kit. Store at 4 °C.
3. Supplemented Dulbecco's Modified Eagle Medium (DMEM) medium (*see* **Note 1**).
4. Supplemented cell culture medium (for immune cells).
5. Gel culture medium: In a 50 mL tube, add 0.5 mL of calcium-based cross-linking solution to 49.5 mL of supplemented DMEM culture medium (5 mM $CaCl_2$ final concentration).
6. Trypan blue solution.

7. De-crosslinking solution: Sterile and ready-to-use from 3D-flowGel kit. Store at 4 °C.
8. Carboxyfluorescein diacetate succinimidyl ester (CFDA-SE).
9. Paraformaldehyde (PFA) supplemented with 1 mM $CaCl_2$: Add 100 μL of 0.5 M $CaCl_2$ to 49.9 mL of PFA.
10. Washing solution: Physiological solution supplemented with 5 mM $CaCl_2$: Add 500 μL of 0.5 M $CaCl_2$ to 49.5 mL of physiological solution.
11. Tumour-antigen-specific primary antibody.
12. Fluorescent-labelled secondary antibody.
13. Fluorescent conjugated immune cell–specific primary antibody for flow cytometry (i.e., PE-anti-CD45).
14. FITC annexin V.

2.2 Material

1. 15 mL conical tube
2. Cell counting chamber.
3. Round bottom 96-well plate.
4. 1 mL syringe
5. 5 mL syringe
6. 2× 21G needle
7. 2 × 6 cm petri dishes
8. MIVO® Single Flow devices with their lids.
9. Inserts with 8.0 μm pore size.
10. Sterile tweezers.
11. 24-well plate.

2.3 Equipment

1. Centrifuge.
2. Humified incubator at 37 °C with 5% CO_2.
3. Peristaltic pump (*see* **Note 2**) with pumping tubes (*see* **Note 3**).
4. Tray.
5. Biological cabinet.
6. Water bath.
7. Inverted microscope.
8. Confocal microscope.
9. Cytofluorimeter.

3 Methods

Carry out all the procedures under the biological cabinet to maintain sterility and at room temperature, unless otherwise specified.

3.1 Immune Cell Preparation

Immune cells need to be prepared at least 48 h before starting the dynamic experiment. Total peripheral blood mononuclear cells (PBMCs) or isolated populations of natural killer (NK), T or B cells, monocytes, or others can be used, based on the specific application.

1. Prepare 5 mL of supplemented cell culture medium in a 15 mL conical tube.
2. Thaw immune cells by immersing the cryovial in a pre-heated 37 °C water bath for about 1–2 min.
3. Transfer the immune cell suspension from the thawed cryovial to the medium-containing tube and slowly pipette.
4. Centrifuge cells at 300 rcf for 7 min and resuspend them in 10 mL of supplemented cell culture medium.
5. Perform a viable cell count using trypan blue solution (1:1 V/V).
6. Adjust the volume of the immune cell suspension to have a final concentration of 1×10^6 viable cells/mL.
7. Transfer 200 μL of immune cell suspension to the round bottom 96-multiwell plate's wells (200,000 cells/well).
8. Move your plate into a humified incubator, at 37 °C and 5% CO_2 and check cell viability every day by performing a viable cell count.

3.2 Three-Dimensional Cancer Model Preparation

The 3D cancer model needs to be prepared 24 h before starting the dynamic experiment. Human cancer cells, either primary cells or cell lines, are embedded in a 3D alginate-based hydrogel to provide a more physiological 3D environment.

1. Culture desired cancer cells in adhesion until their confluence.
2. Detach cells with enzymatic treatment from the plastic surface, based on the specific protocol, and resuspend them in DMEM culture medium, to a final concentration of 2×10^6 cells/mL (*see* **Note 4**).
3. Put 5 mL of the calcium-based cross-linking solution in a 6 cm petri dish, to obtain a cross-linking bath.
4. Put 5 mL of buffer solution in another 6 cm petri dish, to obtain a washing bath.
5. Add 1 mL of alginate solution to 1 mL of cell suspension (1:1 V/V) and mix well by slowly pipetting (*see* **Note 5**).

6. Using a 1 mL syringe and a 21G needle (*see* **Note 6**), aspirate the mixed-up cell solution.
7. Gently drop the cellular solution into the cross-linking bath (*see* **Note 7**) and let them for 1 min to allow the development of the cross-linking reaction and the complete gelification.
8. Using sterile tweezers, gently pick up the gels, one by one, and transfer them into the washing bath.
9. Transfer gels in a porous permeable insert with 8 μm pore size, to support cell passage, and move the insert into the well of a 24-well plate.
10. Move the 24-well plate into a humified incubator, at 37 °C and 5% CO_2.

3.3 Dynamic Co-culture

The dynamic co-culture is performed by using the MIVO® Single Flow device, with the aim to recapitulate the complexity of a 3D, dynamic tumour microenvironment (Fig. 1).

1. Unbox one MIVO® chamber for each experimental replicate under the biological cabinet.
2. Place each MIVO® device on the tray.
3. Close the circuit for every MIVO® device using sterilized pumping tubes.
4. Move the 3D tumour models containing inserts into the MIVO® chamber and fill the insert with 200 μL of gel culture medium.
5. Collect the appropriate number of immune cells from the round bottom 96-well plate into a tube, to get a ratio of effector cells (immune cells): target cells (cancer cells) of 10:1 in every replicate.
6. Centrifuge immune cells at 300 rcf for 7 min.
7. Stain immune cells with 1 μM CFDA-SE for 30 min at 37 °C (*see* **Note 8**).
8. Centrifuge immune cell at 300 rcf for 7 min, to remove the dye excess, and resuspend them in the appropriate volume of gel culture medium (supplemented with specific cytokines or factors, based on the selected cell type).
9. Fill the MIVO® circuit with the immune cells suspension by using a syringe inserted into the three-way valve.
10. Close each MIVO® device with the lid, and move the tray onto the peristaltic pump, within the incubator at 37 °C and 5% CO_2 incubator.
11. Mount the pumping tube on the cartridges of the peristaltic pump.
12. Set the appropriate rpm to recapitulate the capillary flow rate and press "start".

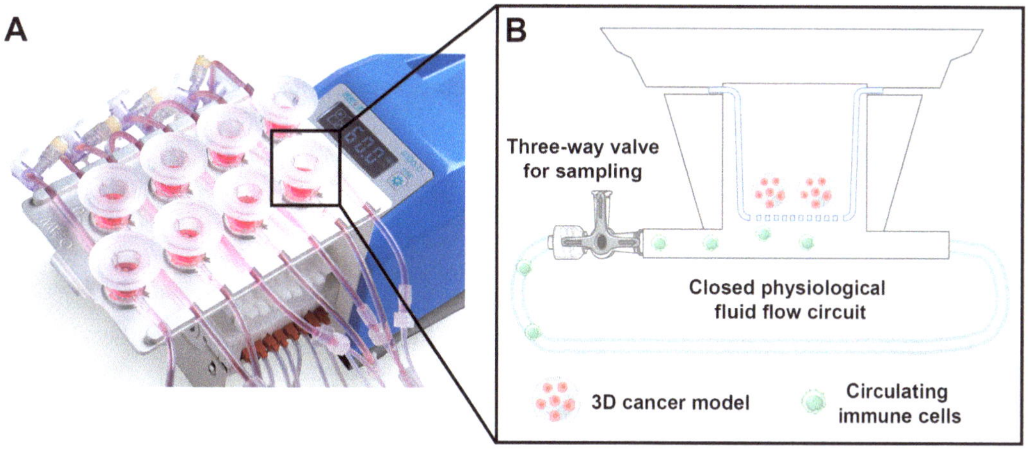

Fig. 1 (**a**) MIVO® fluidic devices with the peristaltic pump. (**b**) Schematic representation of the experimental setting, where the 3D cancer model is cultured in the MIVO® fluidic device, and NK cells are put directly into circulation in the fluid flow, to mimic the bloodstream, under physiological flow conditions

3.4 Analysis of Immune Cells' Extravasation

Immune cells can extravasate from the circulation, to reach the tumour compartment. The extravasation can be evaluated at different timepoints.

1. After 4 h, sample a volume both from the medium in circulation (immune cells compartment) and from the insert containing the 3D tumour model (tumour compartment). Repeat this step in different chambers, to have the proper number of experimental replicates.
2. To determine the percentage of extravasation, count the immune cells present both in the tumour compartment and in the immune cells compartment of every replicate, performing a viable cell count using the trypan blue solution.
3. Repeat **steps 2** and **3** at different time points to assess the kinetic of immune cell extravasation.

3.5 Analysis of Immune Cells' Infiltration

The second step in the immunotherapy process is the infiltration of immune cells into the TME. Immune cells infiltration is analysed by confocal microscopy. The imaging is based on the labelling of immune cells with CFDA-SE, while cancer cells can be stained by using a primary antibody, recognising an antigen specific to the selected tumour model. Perform the following steps in the dark, to preserve fluorochromes.

1. At the end of the experiment, transfer the inserts containing the tumour gel into a 24-well plate, by using tweezers.
2. Wash gels with washing solution and incubate with PFA supplemented with 1 mM $CaCl_2$ for 10 min at room temperature (*see* **Note 9**), for cell fixing.

3. After the fixation step, wash gels twice with washing solution.
4. Incubate gels with tumour-antigen-specific primary antibody for 2 h at room temperature, or at 4 °C overnight (*see* **Note 10**).
5. Wash gels twice with washing solution and incubate them with fluorescent-labelled secondary antibody, for 1 h at room temperature (*see* **Note 11**).
6. Observe at confocal microscope.

3.6 Analysis of Immune Cell-Mediated Tumour Apoptosis

The final step in the immunotherapy assessment is the ability of immune cells to induce apoptosis in cancer cells. Apoptosis can be assessed by evaluating the binding of annexin V on the surface of cancer cells.

1. Using the tweezers, take the gels and transfer them into a 15 mL conical tube (*see* **Note 12**).
2. Add 100 μL of de-crosslinking solution for every collected gel (*see* **Note 13**).
3. Put the tube in the 37 °C water bath.
4. Wait about 15 min for the complete alginate de-crosslinking.
5. Centrifuge the solution at 300 rcf for 7 min.
6. Wash the pellet twice with cell culture medium.
7. Stain the cells with appropriate antibodies, such as anti-CD45 (*see* **Note 14**) to identify immune cells and annexin V (*see* **Note 15**) to identify apoptotic cells, or other specific markers based on the selected model.
8. Analyse the samples by flow cytometry.

The presented experimental setting was used by the authors to recapitulate the immune cell natural killer (NK) migration, infiltration within a 3D tumour matrix, and activation against neuroblastoma cancer cells in a humanized, fluid-dynamic environment. Circulating NK cells actively initiates a spontaneous "extravasation" process towards the physically separated tumour niche, retaining their ability to interact with matrix-embedded tumour cells and to display a cytotoxic effect (tumour cell apoptosis). Since NK cell infiltration and phenotype are correlated with prognosis and response to immunotherapy, their phenotype was also investigated: a clear decrease in CD16-positive NK cells within the migrated and infiltrated population was observed (Fig. 2 [24]).

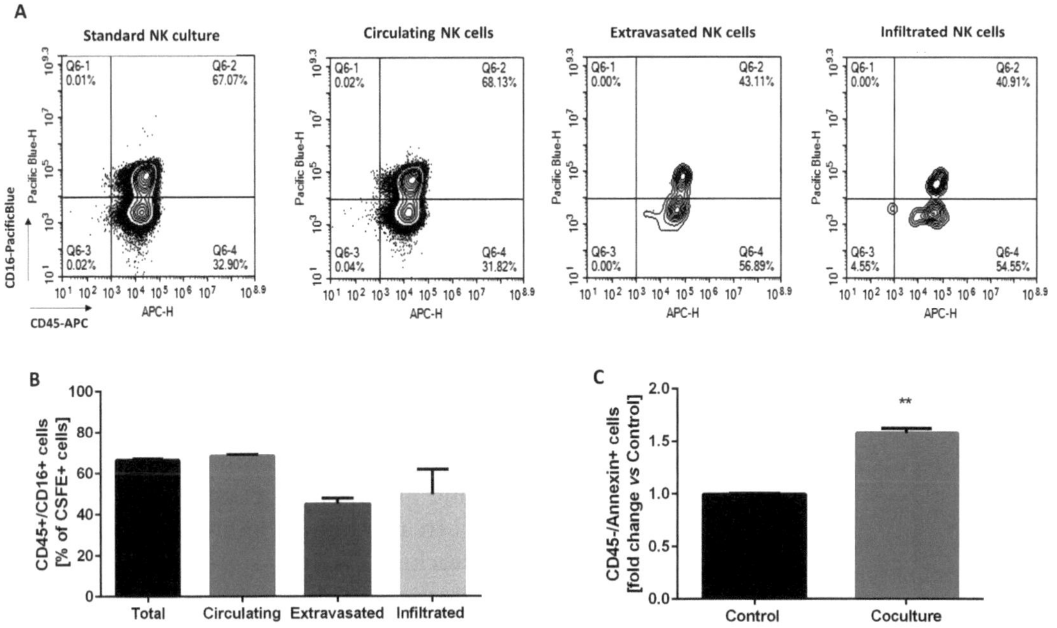

Fig. 2 Phenotypic characterization of the different fractions of NK cells. (**a**) Representative plots showing the phenotypic characterization of the different fractions of NK cells recovered from the dynamic culture (infiltrated, extravasated, circulating vs. the standard NK culture) based on CD45/CD16 expression, gated on the CFSE-positive population ($n = 6$). (**b**) Statistical analysis of CD16 expression in the different fractions of NK cells, recovered from the dynamic culture ($n = 6$). (**c**) Induction of apoptosis in tumour cells, assessed through flow cytometry as annexin V staining on CD45-negative cells ($n = 6$). Control: HTLA-230 3D monoculture. (Copyright © 2022 Marzagalli, Pelizzoni, Fedi, Vitale, Fontana, Bruno, Poggi, Dondero, Aiello, Castriconi, Bottino, and Scaglione. Creative Commons Attribution License (CC BY))

4 Notes

1. The DMEM medium is specifically used to obtain stable gels as it has a low phosphate concentration.
2. Place the peristaltic pump switched on inside the incubator at least 4 h prior to the beginning of the experiment.
3. Sterilize pump tubes in advance by autoclave.
4. A concentration of 2×10^6 cells/ml has been used to reach a final density of about 20,000 cells for each gel, but the cell concentration can be properly adjusted according to the experimental needs.
5. Make sure to not create bubbles when you pipette because they will remain entrapped in the alginate solution and thus in the gels.
6. Using a syringe with the 21G needle, every drop is made up of about 10–20 μL of mixed-up solution. If you want a different

volume, you can either change the needle size or use a pipette setting the exact volume of every drop.

7. Pay attention to keep a proper distance between each drop to allow proper gelification of multiple and distinct gels.
8. CFDA-SE is a highly cell-permeable, non-fluorescent dye, which enters the cytoplasm of cells, where intracellular esterases remove the acetate groups and convert the molecule to the fluorescent ester CFSE.
9. Optimal fixation conditions should be determined individually for each experiment.
10. Optimal antibody dilutions and incubation conditions should be determined individually for each experiment.
11. Be careful to perform secondary antibody dilution and incubation in the dark, paying attention to wrap the consumables used in aluminium foil to avoid direct light exposure that can lead to fluorophore bleaching.
12. Pay attention to gently push all the gels to the bottom of the conical tube.
13. For instance, if you want to dissolve ten gels, you must use 1 mL of alginate-solubilizing solution.
14. CD45 is a marker identifying all the cells derived from the haematopoietic lineage, including immune cells.
15. Annexin V is commonly used to detect apoptotic cells thanks to its ability to bind phosphatidylserine, which is exposed on the outer leaflet of the plasma membrane, during the apoptotic process.

References

1. Stratton MR, Campbell PJ, Futreal PA (2009) The cancer genome. Nature 458:719–724. https://doi.org/10.1038/NATURE07943
2. Vesely MD, Kershaw MH, Schreiber RD, Smyth MJ (2011) Natural innate and adaptive immunity to cancer. Annu Rev Immunol 29: 235–271. https://doi.org/10.1146/ANNUREV-IMMUNOL-031210-101324
3. Rabinovich GA, Gabrilovich D, Sotomayor EM (2007) Immunosuppressive strategies that are mediated by tumor cells. Annu Rev Immunol 25:267–296. https://doi.org/10.1146/ANNUREV.IMMUNOL.25.022106.141609
4. Smyth MJ, Dunn GP, Schreiber RD (2006) Cancer immunosurveillance and immunoediting: the roles of immunity in suppressing tumor development and shaping tumor immunogenicity. Adv Immunol 90:1–50. https://doi.org/10.1016/S0065-2776(06)90001-7
5. Zhang Y, Zhang Z (2020) The history and advances in cancer immunotherapy: understanding the characteristics of tumor-infiltrating immune cells and their therapeutic implications. Cell Mol Immunol 17:807–821. https://doi.org/10.1038/S41423-020-0488-6
6. Demaria O, Cornen S, Daëron M et al (2019) Harnessing innate immunity in cancer therapy. Nature 574:45–56. https://doi.org/10.1038/S41586-019-1593-5
7. Woo SR, Corrales L, Gajewski TF (2015) Innate immune recognition of cancer. Annu Rev Immunol 33:445–474. https://doi.org/10.1146/ANNUREV-IMMUNOL-032414-112043

8. Anderson NM, Simon MC (2020) The tumor microenvironment. Curr Biol 30:R921–R925. https://doi.org/10.1016/J.CUB.2020.06.081
9. Boucherit N, Gorvel L, Olive D (2020) 3D tumor models and their use for the testing of immunotherapies. Front Immunol 11. https://doi.org/10.3389/FIMMU.2020.603640
10. Ando Y, Mariano C, Shen K (2021) Engineered in vitro tumor models for cell-based immunotherapy. Acta Biomater 132:345–359. https://doi.org/10.1016/J.ACTBIO.2021.03.076
11. Katt ME, Placone AL, Wong AD et al (2016) In vitro tumor models: advantages, disadvantages, variables, and selecting the right platform. Front Bioeng Biotechnol 4. https://doi.org/10.3389/FBIOE.2016.00012
12. Amatangelo MD, Bassi DE, Klein-Szanto AJP, Cukierman E (2005) Stroma-derived three-dimensional matrices are necessary and sufficient to promote desmoplastic differentiation of normal fibroblasts. Am J Pathol 167:475–488. https://doi.org/10.1016/S0002-9440(10)62991-4
13. Cassidy JW, Caldas C, Bruna A (2015) Maintaining tumor heterogeneity in patient-derived tumor xenografts. Cancer Res 75:2963–2968. https://doi.org/10.1158/0008-5472.CAN-15-0727
14. Decker WK, da Silva RF, Sanabria MH et al (2017) Cancer immunotherapy: historical perspective of a clinical revolution and emerging preclinical animal models. Front Immunol 8. https://doi.org/10.3389/FIMMU.2017.00829
15. Mestas J, Hughes CCW (2004) Of mice and not men: differences between mouse and human immunology. J Immunol 172:2731–2738. https://doi.org/10.4049/JIMMUNOL.172.5.2731
16. Morgan KM, Riedlinger GM, Rosenfeld J et al (2017) patient-derived xenograft models of non-small cell lung cancer and their potential utility in personalized medicine. Front Oncol 7. https://doi.org/10.3389/FONC.2017.00002
17. Vitale C, Marzagalli M, Scaglione S et al (2022) Tumor microenvironment and hydrogel-based 3D cancer models for in vitro testing immunotherapies. Cancers (Basel) 14:1013. https://doi.org/10.3390/CANCERS14041013
18. Rodrigues J, Heinrich MA, Teixeira LM, Prakash J (2021) 3D In vitro model (r)evolution: unveiling tumor-stroma interactions. Trends Cancer 7:249–264. https://doi.org/10.1016/J.TRECAN.2020.10.009
19. Esch EW, Bahinski A, Huh D (2015) Organs-on-chips at the frontiers of drug discovery. Nat Rev Drug Discov 14:248–260. https://doi.org/10.1038/NRD4539
20. Zhang B, Korolj A, Lai BFL, Radisic M (2018) Advances in organ-on-a-chip engineering. Nat Rev Mater 3(8):257–278. https://doi.org/10.1038/s41578-018-0034-7
21. Fischbach C, Chen R, Matsumoto T et al (2007) Engineering tumors with 3D scaffolds. Nat Methods 4:855–860. https://doi.org/10.1038/nmeth1085
22. Görlach A, Herter P, Hentschel H et al (1994) Effects of nIFN β and rifn γ on growth and morphology of two human melanoma cell lines: comparison between two-and three-dimensional culture. Int J Cancer 56:249–254. https://doi.org/10.1002/ijc.2910560218
23. Wikswo JP (2014) The relevance and potential roles of microphysiological systems in biology and medicine. Exp Biol Med 239:1061–1072. https://doi.org/10.1177/1535370214542068
24. Marzagalli M, Pelizzoni G, Fedi A et al (2022) A multi-organ-on-chip to recapitulate the infiltration and the cytotoxic activity of circulating NK cells in 3D matrix-based tumor model. Front Bioeng Biotechnol 10. https://doi.org/10.3389/fbioe.2022.945149
25. Marrella A, Fedi A, Varani G et al (2021) High blood flow shear stress values are associated with circulating tumor cells cluster disaggregation in a multi-channel microfluidic device. PLoS One 16:e0245536. https://doi.org/10.1371/journal.pone.0245536
26. Marrella A (2020) 3D fluid-dynamic ovarian cancer model resembling systemic drug administration for efficacy assay. ALTEX. https://doi.org/10.14573/altex.2003131

Chapter 12

Isolation and Single-Cell Transcriptomic Analysis of Murine Regulatory B Cells

Razan M. Alajoleen, Liwu Li, and Xin M. Luo

Abstract

Regulatory B (Breg) cells have been demonstrated to play an important role in the inhibition of a wide range of immunological responses, and they are absent or malfunction in autoimmune diseases like lupus. Breg cells can control immunological responses and keep the immune system in a balanced state by releasing immunosuppressive cytokines such as transforming growth factor-beta (TGF-β) and interleukin-10 (IL-10), which in turn promote regulatory T (Treg) cells and reduce effector T cell responses. Breg cells have also been linked to the modulation of cancer immunity. Due to their immunosuppressive role, in the context of cancer, Breg cells aid in tumor immune evasion and promote tumor progression. Nonetheless, it has been established that Breg cells are involved in both cancer immunity and autoimmunity, and their characterizations beyond surface markers, for example, on the transcriptomic level, are essential for our understanding of Breg biology in health and disease. In this chapter, using lupus-prone MRL/*lpr* mice, we describe a Breg cell isolation protocol for the purpose of single-cell RNA sequencing analysis.

Key words Regulatory B cells, Autoimmune disease, Cancer, Single-cell RNA sequencing

1 Background

Regulatory B (Breg) cells have been known to develop in response to inflammation. They are crucial in controlling a variety of autoimmune diseases [1]. Breg cells are a significant part of the immune system, even though their definition and mechanism of action are still not clear. Breg cells are heterogeneous in their phenotype, but they have the common capacity to produce immunosuppressive cytokines such as interleukin-10 (IL-10). Systemic lupus erythematosus (SLE) is a multisystemic disorder characterized by a high level of autoantibody production and immune complex deposition. Multiple organ damage, including lesions in the skin and kidneys, is a hallmark of SLE. The development of SLE is influenced by the imbalance of effector T cells, alterations in the production of associated cytokines, and overactivation of T helper-2 (Th2) cells [2].

Vijay Kumar (ed.), *Immune Homeostasis: Methods and Protocols*, Methods in Molecular Biology, vol. 2782,
https://doi.org/10.1007/978-1-0716-3754-8_12,

B cell activation and differentiation are mediated by T helper cells [3], and deregulation of these cells can result in the generation of autoantibodies [3]. Additionally, Th1 and Th2 cells can actively cause tissue damage by releasing inflammatory cytokines [4]. The regulation of antibody-producing B cells is disrupted in SLE, and studies have shown that SLE patients have increased numbers of autoantibody-producing B cells, including plasma cells and memory B cells [5, 6].SLE has also been linked to the pathophysiology of double-negative (DN-T) cells, namely T cells lacking the expression of CD4 and CD8 co-receptors, which contribute to the development of autoimmunity by producing pro-inflammatory cytokines and activating autoreactive B cells [7].

Recently, growing evidence suggests that IL-10 can control the onset and progression of autoimmune disorders [8]. Cytokine release from effector T cells, for example, is inhibited by IL-10 [9]. IL-10 can also enhance regulatory T (Treg) cell-mediated tolerance to the peripheral environment by triggering the STAT3 pathway [10]. Additionally, IL-10 controls antigen presentation, inflammation, and the generation of autoantibodies [11]. It also affects B cell activation and death [12]. Importantly, Breg cells are known to produce the signature cytokine IL-10 [13–15], which is commonly used as a phenotypic marker for Breg cells. Due to the heterogenicity of Breg cells, we define them as any B cells that can produce IL-10 upon stimulation with lipopolysaccharide (LPS). Using single-cell RNA sequencing (scRNA-seq), we aimed to characterize the transcriptomic profiles of diverse Breg subpopulations present in the spleen of lupus-prone MRL/*lpr* mice.

2 Materials

Prepare all solutions in a sterile hood. Prepare and store all reagents at 4 °C. Prepare all solutions using sterile 1X phosphate-buffered saline (PBS).

1. Spleens from Murphy Roths Large/Lymphoproliferation (MRL/*lpr*) female mice at 6 weeks of age (body weight around 36 g) were used for the isolation of Breg cells (*See* **Note 1**). Mice were purchased from The Jackson Laboratory (stock number 000485) and bred and maintained in a specific pathogen-free facility according to the requirements of the Institutional Animal Care and Use Committee (IACUC) at Virginia Tech.
2. 5 mL syringes (BD 309575, USA).
3. 70 μm cell strainers (Fisher Scientific, 22363548, USA).
4. 15 mL and 50 mL conical tubes (Fisher Scientific, USA).

5. Complete RPMI (Roswell Park Memorial Institute) (C10 medium): RPMI-1640, 10% fetal bovine serum (FBS), 1 mM sodium pyruvate, 1% 100× MEM nonessential amino acids, 10 mM HEPES, 55 μM 2-mercaptoethanol, 2 mML-glutamine, 100 U/ml penicillin–streptomycin (Life Technologies, Grand Island, NY).
6. Magnetic activated cell sorting (MACS) Buffer: 2% FBS, 2 mM EDTA in 1X PBS (Life Technologies, USA).
7. Miltenyi Biotec Regulatory B Cell Isolation Kit, mouse (Cat #130–095-873, USA).
8. Red blood cell (RBC) lysis buffer (Biolegend, 420301, USA).
9. Lipopolysaccharides (LPS) from *Salmonella enterica* serotype enteritidis (*see* **Note 2**). (Sigma, L7770, USA).
10. Cell stimulation cocktail (500X)-containing PMA/Ionomycin (ThermoFisher, 00-4975-03, USA).
11. Fluorochrome-conjugated anti-mouse CD19 (Biolegend, USA).
12. Propidium iodide (PI) (Fisher Scientific-00-6990-50, USA).
13. Fluorescence-activated cell sorting (FACS) buffer: 5% fetal bovine serum (FBS) in 1X phosphate buffer saline (PBS) (Life Technologies, USA).
14. 5 mL round bottom polystyrene FACS tubes with caps (Fisher Scientific, FSC-T9010, USA).
15. BD FACS Aria fusion cell sorter (BD Biosciences, San Jose, CA).
16. Chromium Next GEM Single Cell 3′ GEM, Library & Gel Bead Kit v3.1 (10X Genomics, PN-1000128, USA).

3 Methods

The B cell enrichment protocol is for one female MRL/*lpr* mouse. Scale up volumes accordingly. All cell suspensions and associated buffers should be kept ice cold for cell isolation unless stated otherwise.

1. Euthanize the mouse by CO_2 inhalation.
2. Pin the mouse to the dissection plate with the ventral side facing up. Use 70% ethanol to spray the incision area.
3. Cut the skin along the ventral midline, from the pubic symphysis to the neck, and separate it from the muscle wall beneath. Pin the skin to both sides of the body.
4. Cut the muscle layer and locate the spleen next to the stomach and beneath the small intestine on the right side of the body. Take the spleen by one end and pull it away from the stomach.

5. Harvest the spleen from the mouse, keep on ice in 5 mL MACS buffer in a 15 conical tube.
6. Empty the tube containing spleen on a 70 μm strainer docked on a 50 mL conical tube.
7. Mash the spleen with a syringe plunger for 30–60 s.
8. Wash the strainer with 10 mL MACS buffer.
9. Remove the strainer, cap the conical tube containing single cell suspension and centrifuge at 350× *g* for 5 min at 4 °C.
10. Remove supernatant and resuspend pellet in 1 mL 1X RBC lysis buffer for 5 min in the dark at room temperature.
11. Add 4 mL MACS buffer and centrifuge at 350× *g* for 5 min at 4 °C.
12. Remove supernatant and resuspend pellet in 5 mL MACS buffer and count the cells.
13. Negatively isolate B cells using immunomagnetic separation technology (*see* **Note 3**). Follow the instructions of the kit.
14. Count isolated cells and resuspend in complete RPMI. Analyze the enriched B cells using FACS (Fig. 1).
15. Seed cells at 2.5×10^6 cells per well using a 12-well plate and stimulate the cells with 10 μg/mL LPS. Once the cells have been seeded, put the plate in the 37 °C incubator for 24 h. Five hours before the end of incubation, add cell stimulation cocktail as recommended by the manufacturer.

The following are the Breg cell enrichment protocol and scRNA seq.

1. From the 12-well plate, collect the cells in a 15 mL tube and count the cells.
2. Isolate Breg cells using Regulatory B Cell Isolation Kit, mouse. Follow the instructions of the kit.
3. Resuspend cells at 5×10^6 per 3 mL in FACS buffer and centrifuge at 350 × *g* for 5 min at 4 °C.
4. Prepare anti-mouse CD16/32 antibody solution (1:50 dilution in FACS buffer, 5 μg/mL final concentration). Resuspend the cell pellet in 100 μL of anti-mouse CD16/32 antibody solution for 10 min on ice. Then add 5 mL of FACS buffer and centrifuge at 350× *g* for 5 min at 4 °C.
5. Prepare APC (Allophycocyanin)-conjugated anti-mouse CD19 (1:800 dilution in FACS buffer) as recommended by the manufacturer at 100 μL final volume per sample.
6. After centrifugation, aspirate the supernatant and resuspend the cell pellet in 100 μL anti-mouse CD19. Incubate on ice for 30 min in the dark.

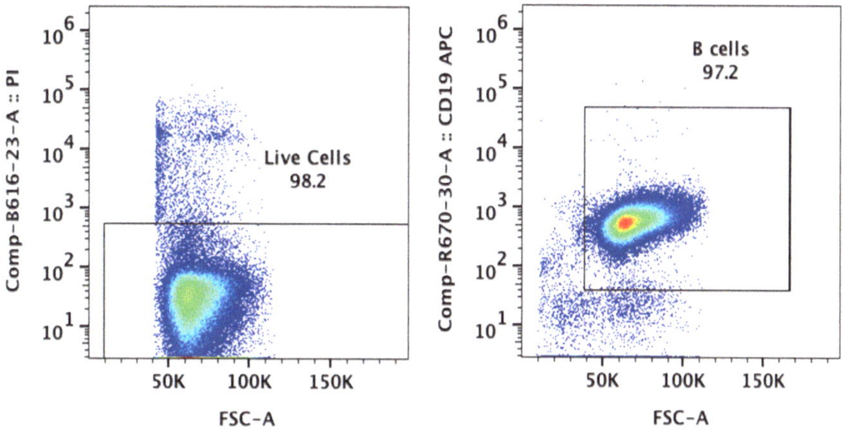

Fig. 1 Purity plots following B cell enrichment

7. Add 5 μL of PI staining solution per 100 μL of cells. Incubate on ice for 5 min in the dark. Then add 5 mL FACS buffer and centrifuge at 350× *g* for 5 min at 4 °C.
8. After centrifugation, aspirate the supernatant and resuspend the cell pellet in 100 μL ice-cold FACS buffer. Keep the cells on ice until in the dark until sorting. Sort within 1 h.
9. Single color staining:
   (a) Label 3 FACS tubes with the APC, PI, or unstained. Aliquot cells into the FACS tubes at 1 × 10^6 cells per tube, followed by a 4 mL FACS wash at 800× *g* for 5 min at 4 °C.
   (b) Resuspend the cell pellet in 100 uL FACS buffer after aspirating the supernatant.
   (c) Add one of the following antibodies to each corresponding FACS tube: anti-mouse CD19-APC 1: 800 dilution using FACS buffer; PI 1:20 dilution using 1× PBS at 100 μL final volume per sample; for the unstained sample, add 100 μL FACS buffer only.
   (d) Let them sit on ice in the dark for 30 min.
   (e) Add 3 mL of ice-cold FACS to each tube after incubation, and then centrifuge the FACS tubes at 350× *g* for 5 min at 4 °C.
   (f) After centrifugation, remove the supernatant and resuspend all cell pellets in 100 μL of ice-cold FACS buffer. All three tubes should be kept in the dark and on ice until sorting.
10. Set the gates of the sorted cell populations by using single stained samples from 3.24. Gating the target cell population manually results in a cluster of dots that appear to be distinct

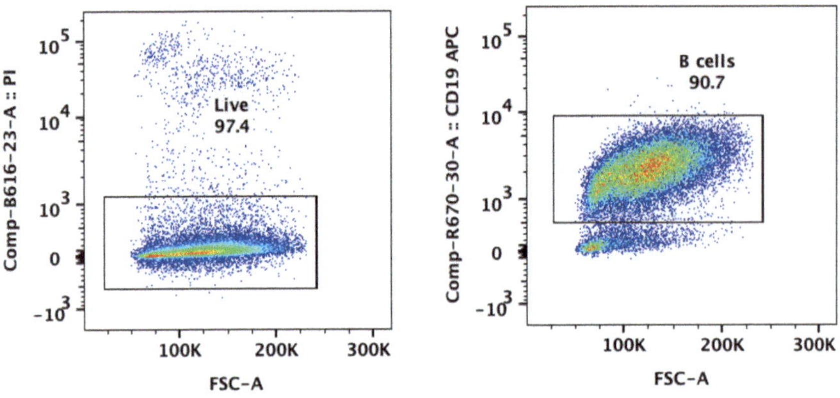

Fig. 2 Gating strategy for Breg cell sorting. Enriched Breg cells after 24 h stimulation with 10 μg/mL LPS were stained with antibodies against CD19 and PI for live/dead cells for FACS sorting

from other dots using fluorescence intensity as parameters of the X- and Y-axis graph. Note: Depending on the needs of the researchers, this gating set is adjustable and arbitrary. Move the gate higher based on the matching fluorescence intensity if researchers anticipate higher purity based on one or more markers [16].

11. Prepare collection FACS tubes with 15 μL 50% FBS Complete RPMI media.
    (a) Fill the collection tube with the target sorted cell population (Live $PI^{-}CD19^{+}$ cells; Fig. 2). Sort all the prepared cells.
    (b) Fill the collection tube with 4 mL of ice-cold MACS buffer, cap it, and invert it to mix.
    (c) Centrifuge the collecting tube at 350× *g* for 5 min at 4 °C, after which aspirate the supernatant, leaving around 200 μL of the cell pellet.
    (d) Resuspend the cell pellet with 0.2 mL of ice-cold MACS buffer.
    (e) After manual counting, 10,000 Breg cells are loaded for capture onto the Chromium System using the Chromium Next GEM Single Cell 3′ GEM, Library & Gel Bead Kit v3.1 (10X Genomics).
    (f) Following capture and lysis, cDNA is synthesized and amplified (12 cycles) as per the manufacturer's protocol (10X Genomics).
    (g) Run the sample (Dilution Factor 1:10) on an Agilent Bioanalyzer high-sensitivity chip. TapeStation test (quality control) is used to evaluate the quality of the cDNA libraries prior to sequencing. The representative TapeStation result is shown in Fig. 3.

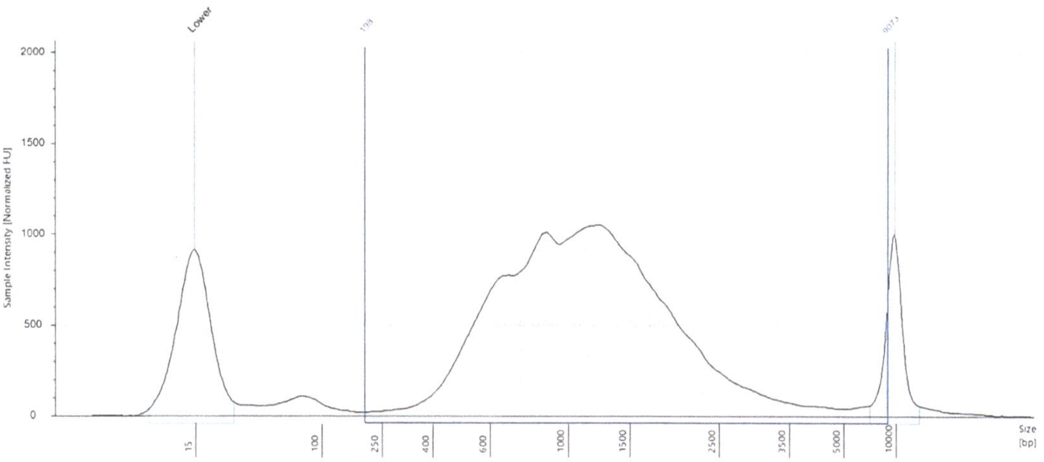

Fig. 3 Electropherogram pattern of cDNA, showing maximum peak size between 600 and 1500 bp cDNA separated with the D5000 ScreenTape assay and the TApeStation system

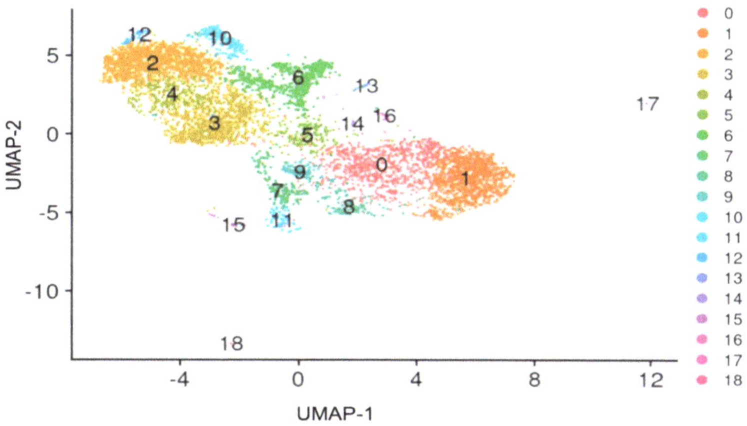

Fig. 4 UMAP plot showing 19 clusters across 10,000 Breg cells from an MRL/*lpr* female mouse

(h) The amplified cDNA is used to construct an Illumina sequencing library and sequenced on a HiSeq 4000.

(i) Sequencing data were analyzed using Surat software and visualized using Loupe Browser (Fig. 4). For detailed product-specific information, visit the 10X Genomics Support website.

4 Notes

1. MRL/*lpr* mice can spontaneously develop clinical signs similar to those of human SLE and are a classic model of the disease.
2. LPS is used to induce the production of IL-10 from existing Breg cells, allowing the isolation of Breg cells.

3. Several different B cell enrichment kits can be used. Since 100% purity is not required, reagents can be diluted more than advised. The main goal of this stage is to enrich B cells to speed up the cell sorting rather than sorting B cells directly from splenocytes. However, we advise utilizing a negative selection kit for B cell enrichment.

References

1. Matsushita T (2019) Regulatory and effector B cells: friends or foes? J Dermatol Sci 93(1):2–7. https://doi.org/10.1016/j.jdermsci.2018.11.008
2. Wang T, Mei Y, Li Z (2019) Research progress on regulatory B cells in systemic lupus erythematosus. Biomed Res Int 2019:7948687. https://doi.org/10.1155/2019/7948687
3. Janeway CA Jr, Travers P, Walport M et al (2001) B-cell activation by armed helper T cells. In: Immunobiology: the immune system in health and disease, 5th edn
4. Berger A (2000) Th1 and Th2 responses: what are they? BMJ 321(7258):424. https://doi.org/10.1136/bmj.321.7258.424
5. Odendahl M, Jacobi A, Hansen A, Feist E, Hiepe F, Burmester GR et al (2000) Disturbed peripheral B lymphocyte homeostasis in systemic lupus erythematosus. J Immunol 165(10):5970–5979. https://doi.org/10.4049/jimmunol.165.10.5970
6. Wei C, Anolik J, Cappione A, Zheng B, Pugh-Bernard A, Brooks J et al (2007) A new population of cells lacking expression of CD27 represents a notable component of the B cell memory compartment in systemic lupus erythematosus. J Immunol 178(10):6624–6633. https://doi.org/10.4049/jimmunol.178.10.6624
7. Crispín JC, Oukka M, Bayliss G, Cohen RA, Van Beek CA, Stillman IE et al (2008) Expanded double negative T cells in patients with systemic lupus erythematosus produce IL-17 and infiltrate the kidneys. J Immunol 181(12):8761–8766. https://doi.org/10.4049/jimmunol.181.12.8761
8. Gibson AWEJ, Wu J et al (2000-2013) The role of IL-10 in autoimmune pathology. In: Madame curie bioscience database [Internet]
9. Moore KW, de Waal MR, Coffman RL, O'Garra A (2001) Interleukin-10 and the interleukin-10 receptor. Annu Rev Immunol 19:683–765. https://doi.org/10.1146/annurev.immunol.19.1.683
10. Chaudhry A, Samstein RM, Treuting P, Liang Y, Pils MC, Heinrich JM et al (2011) Interleukin-10 signaling in regulatory T cells is required for suppression of Th17 cell-mediated inflammation. Immunity 34(4):566–578. https://doi.org/10.1016/j.immuni.2011.03.018
11. Mittal SK, Roche PA (2015) Suppression of antigen presentation by IL-10. Curr Opin Immunol 34:22–27. https://doi.org/10.1016/j.coi.2014.12.009
12. Itoh K, Hirohata S (1995) The role of IL-10 in human B cell activation, proliferation, and differentiation. J Immunol 154(9):4341–4350
13. Fillatreau S, Sweenie CH, McGeachy MJ, Gray D, Anderton SM (2002) B cells regulate autoimmunity by provision of IL-10. Nat Immunol 3(10):944–950. https://doi.org/10.1038/ni833
14. Mizoguchi A, Mizoguchi E, Takedatsu H, Blumberg RS, Bhan AK (2002) Chronic intestinal inflammatory condition generates IL-10-producing regulatory B cell subset characterized by CD1d upregulation. Immunity 16(2):219–230. https://doi.org/10.1016/S1074-7613(02)00274-1
15. Iwata Y, Matsushita T, Horikawa M, DiLillo DJ, Yanaba K, Venturi GM et al (2011) Characterization of a rare IL-10–competent B-cell subset in humans that parallels mouse regulatory B10 cells. Blood 117(2):530–541. https://doi.org/10.1182/blood-2010-07-294249
16. Liao X, Makris M, Luo XM (2016) Fluorescence-activated cell sorting for purification of Plasmacytoid dendritic cells from the mouse bone marrow. J Vis Exp 117. https://doi.org/10.3791/54641

Chapter 13

Isolation and Culture of Mouse Primary Microglia and Oligodendrocyte Precursor Cells

Zihao Li and Cun-Jin Zhang

Abstract

Microglia and oligodendrocyte precursor cells (OPCs) are critical glia subsets in the central nervous system (CNS) and are actively engaged in a body of diseases, such as stroke, Alzheimer's disease, multiple sclerosis, etc. Microglia and OPC serve as compelling tools for the study of CNS diseases as well as the repair and damage of myelin sheath in vitro. In this protocol, we summarized a method which is capable of using the same batch of new-born mice to isolate and culture microglia and OPCs. It integrates the characteristics of practicality, convenience, and efficiency, providing a convenient, easy, and reliable technique for research.

Key words Microglia, OPC, Cell culture, Protocol, CNS

1 Introduction

It is universally acknowledged that microglia act as phagocytes responsible for clearing debris during the development of central nervous system and diseases [1]. Indeed, microglia represent a pivotal link between neurological and immunological activities in the CNS [1, 2]. When inflammation, infection. or other neurological diseases occur in the brain, microglia are rapidly activated and obtain phagocytic function [3]. The morphological changes of microglia reflect the activated state, which is closely related to the severity of brain damage. The in vitro model of microglia is also used to study specific molecular pathways so that we can have more profound insights into CNS diseases.

Oligodendrocytes are myelin glial cells in the CNS, which play a significant role in the formation of myelin sheath, supporting axonal and promoting the conduction of action potential [4–6]. Oligodendrocyte precursor cells (OPCs) proliferate and migrate in the CNS at the stage of late embryonic development and further differentiate into mature myelinating oligodendrocyte [4, 7]. Similarly, the isolation and purification of OPCs not only contribute to a

Vijay Kumar (ed.), *Immune Homeostasis: Methods and Protocols*, Methods in Molecular Biology, vol. 2782,
https://doi.org/10.1007/978-1-0716-3754-8_13,

better understanding of oligodendrocyte function and development but also provide a reliable tool for myelin-damage diseases.

Among the existing methods for isolating and culturing microglia, there are different methods for extracting microglia from adult mice and neonatal mice. These methods have their own characteristics. After reviewing a large amount of culture methods for microglia and OPCs [8–11], we summarized the most simple and stable method for isolation and culture of mouse primary microglia and OPCs by using neonatal mice. This protocol has a high yield with simple and common steps and reagents.

2 Materials

This section includes sterilization preparation, material preparation, and reagent configuration.

2.1 Sterilization Preparation

The day before the operation:

1. Wrap the glass petri dishes (two pieces, wrapped in tin foil), micro tweezers (WA3050), ordinary small tweezers, and scissors in tin foil and put them into the box; autoclave sterilization.
2. The instruments in the instrument box should be replaced with new alcohol 1 day before cell culture and soaked overnight.

Before operation:

1. Wipe the microscope with 75% alcohol and then move it to the ultra-clean bench for sterilization.
2. Put tweezers into a small beaker filled with alcohol, and a small beaker filled with dissection solution and put it into the ultra-clean bench.
3. Put the petri dishes (four 10 cm dishes and one 6 cm dish) into the ultra-clean bench. UV sterilize for 30 min.

2.2 Material Preparation

P1-P2 neonatal mice, ice pack, 75% alcohol, 15 mL centrifuge tube, 50 mL centrifuge tube, dissection solution (*See* Subheading 2.3 for details), trypsin-EDTA, FBS, Dulbecco's Modified Eagle Medium: high glucose, DMEM: F12, poly-D-lysine(PDL)syringe, basic fibroblast growth factor (bFGF), penicillin-streptomycin (100×), B-27™ plus(50X)(An optimized serum-free additive used to support the low- or high-density growth and short-term or long-term survival of embryonic, postnatal, and adult hippocampus and other CNS neurons.), platelet-derived growth factor-AA (PDGF-AA), ciliary neurotrophic factor (CNTF), triiodothyronine (T3), filter, sterilized microscope, petri dishes, cell counter, 75 cm^2 flask.

2.3 Reagent Configuration

1. Dissection solution.

   Steps: 1 × PBS 975 mL; add 25 mL 100 × penicillin-streptomycin; filtered by 0.22 μm filter.

2. 10% primary microglia medium (PMG)

   Steps: DMEM high glucose 45 mL; add 5 mL FBS; add 500 μL penicillin-streptomycin (100×) (working concentration 100 ug/mL); filter.

DMEM high glucose	45 mL
Pen/strep(100×)	500 μL
FBS (50X)	5 mL

3. 0.25% Trypsin-EDTA
4. OPC proliferation medium.

DMEM/F12	49 mL
Pen/strep(100×)	250 μL
B-27™ plus (50X)	1 mL
bFGF (10 ng/mL)	5 μL (10 ng/mL)
PDGF-AA (30 ng/mL)	15 μL (30 ng/mL)

   Filter

5. OPC Differentiation Medium

DMEM/F12	49 mL
Pen/strep(100×)	250 μL
B-27™ plus (50X)	1 mL
Cntf	10 μL (20 ng/mL)
T3 (triiodothyronine)	25 μL (30 nM)

   Filter

3 Methods

1. Spray alcohol onto ice packs and put them into the ultra-clean bench (*see* **Note 1**); place a petri dish on it and pour an appropriate amount of DMEM incomplete medium; pour an appropriate amount of dissection solution into three dishes, and pour an appropriate amount of alcohol into the other dish for standby.

2. Transfer the petri dish with the mice (Fig. 1a) to the ultra-clean bench, cut off the head with scissors, soak the head in alcohol for about 20 s, and quickly transfer it to the dissection solution to rinse the residual alcohol for a while (the residual blood can also be washed). Transfer into another dish with dissection solution. Remove the mouse carcass, scissors, and a petri dish containing alcohol and the used dissection solution (*see* **Note 2**).
3. Turn off the light, turn on the microscope, and separate the brain tissue under the microscope.
   (a) Use the first set of forceps to peel off the skin and dura (cranium) layer by layer to expose the structures of the cerebral cortex, cerebellum, and brain stem. Afterwards, the brain tissue was gently opened and transferred to another dish containing PBS (*see* **Note 3**) (Fig. 1b).
   (b) Use the second set of forceps to peel off the olfactory bulb, brain stem, and cerebellum, thereby separating the brain and further removing the pia mater. Dissociated brain tissue was placed in a DMEM-prepared dish on ice (*see* **Note 4**) (Fig. 1c).
4. Check again under the microscope whether the pia mater of the collected brain tissue is completely peeled off and withdraw from the dissecting microscope.
5. Discard DMEM, add an appropriate amount of trypsin (*see* **Note 5**), and place in a 37 °C cell incubator for 10 min.
6. Aspirate the trypsin, add 2–3 mL PMG to stop digestion, gently pipette to form a cell suspension, aspirate the above cell suspension every five to ten times, collect it into a 50 mL centrifuge tube with a 75 μm filter (*see* **Note 6**), and then add 2–3 mL of new medium and continue pipetting.
7. Centrifuge at 1000 rpm, 5 min, 4 °C.
8. Discard the supernatant and add 3–4 mL of 10% PMG medium to resuspend (for microglia); or add 2 mL of OPC proliferation medium to resuspend (for OPCs).

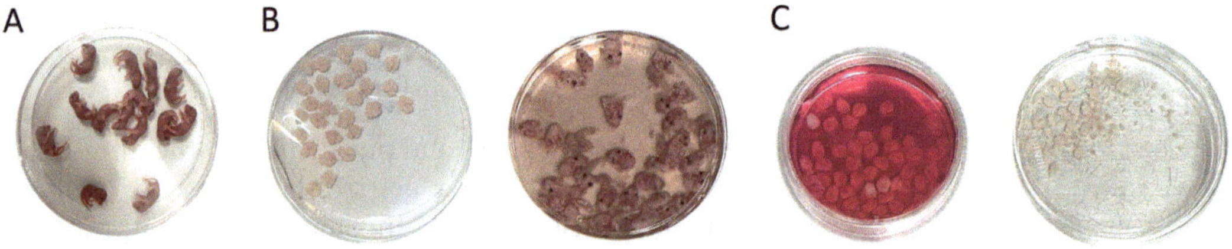

Fig. 1 The process of separating the cerebral cortex. (**a**) P1-P2 neonatal mice; (**b**) The brain tissues which peeled off the skin and dura (cranium); (**c**) Cerebral cortex; removed olfactory bulb, brain stem, cerebellum, and pia mater

9. Cell count (via cell counter or other methods).
10. Cell seed to 75 cm^2 FLASK with 10% PMG 12 ml (about one 75cm^2 FLASK per brain).
11. Put into the 37 °C cell incubator.
12. Change the medium for the first time after 36–48 h.
13. After 6–7 days, change the medium for the second time after the astrocytes cover the bottom of the bottle.

3.1 Microglia Harvest (Fig. 2a, b)

1. The cells were harvested for the first time by shaking the culture flask for about 10 days.
2. After about 13 days of culture, shake the culture flask and collect the cells for the second time.

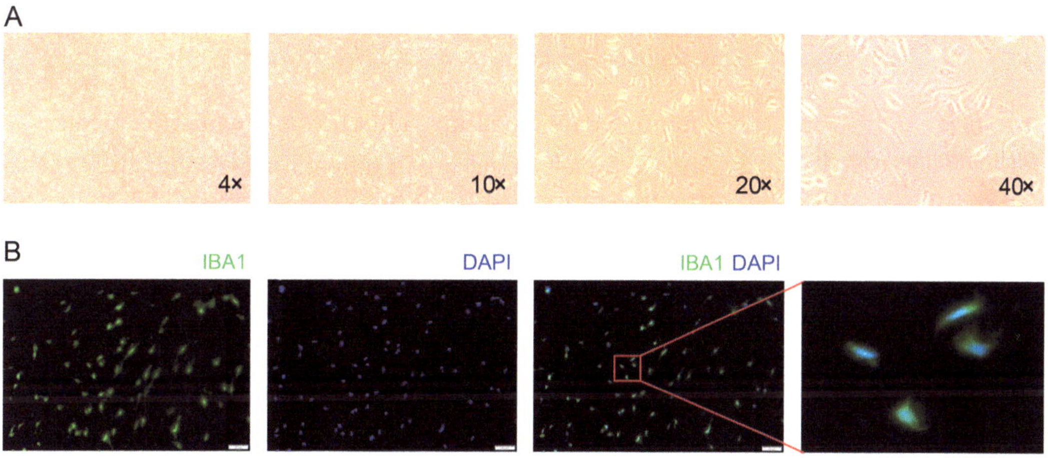

Fig. 2 Primary mouse microglia. (**a**) Microscopic morphology of microglia (4 × , 10 × , 20 × , 40×); (**b**) Immunofluorescence of microglia, scale bar = 50 μm

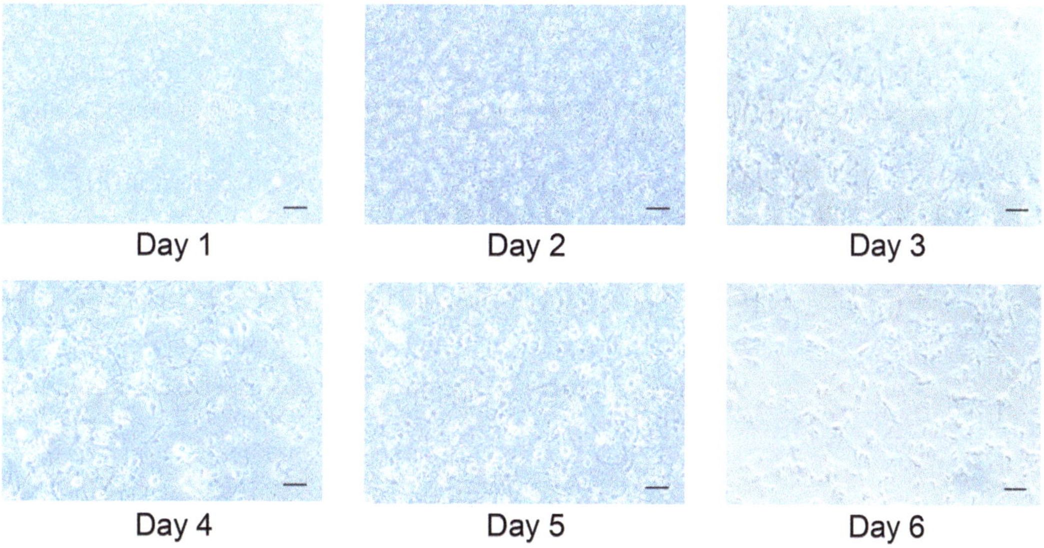

Fig. 3 OPCs cultured in OPC differentiation medium. Scale bar = 200 μm

3.2 OPC Proliferation

1. Collect cells after 4–5 days of culture in OPC proliferation medium and centrifuge at 1000 rpm, 5 min.
2. Transfer the supernatant to a new tube for standby.
3. Use 1 mL of 0.25% trypsin-EDTA-suspended cells, and they are digested for about 2–3 min.
4. Dilute by supernatant collected in step 2 to 8 mL, shake gently, mix well (*see* **Note 7**), and centrifuge at 800 rpm, 3 min.
5. Discard the supernatant and add 2 mL of OPC proliferation medium for resuspension (gently resuspend into single cell suspension).
6. The cells were seeded into the plate coated with poly-D-Lysine according to the amount of cell.
7. Observe the cell state after 3–5 days of culture and change the medium once every 2–3 days (*see* **Note 8**).

3.3 OPC Differentiation (Fig. 3)

1. After 3–5 days of culture, add OPC differentiation medium.
2. After 4–5 days of culture in differentiation medium, reticular structure can appear (*see* **Note 9**).

4 Notes

1. The ultra-clean bench here needs to be sterilized by UV more than half an hour in advance.
2. The suckling mice used in this step should be strictly controlled for P1-P2.
3. During the peeling process, the left hand can be used to insert the straight forceps into the two eyes for fixation, and the right hand can be used to peel off the sagittal suture from front to back.
4. The forceps should be rinsed in dissection solution after contact with alcohol during the operation to avoid direct contact with the brain tissue by alcohol.
5. According to our experience, about 6 mL of trypsin-EDTA for ten brains is appropriate.
6. The pipette tip is added vertically to avoid damage to the cell.
7. Do not resuspend the cells violently.
8. It can be transfected with virus during this period.
9. It can be processed according to your experiment during this period.

Acknowledgement

This study was supported by the National Natural Science Fund for Excellent Young Scholars (82022019 to C.-J. Zhang), Nanjing Special Fund for Medical Science and Technology Development Projects for Distinguished Young Scholars (JQX19005 to C.-J. Zhang), and National Natural Science Foundation of China (81701235 to C.-J. Zhang, 81991514 to C.-J. Zhang).

References

1. Prinz M, Jung S, Priller J (2019) Microglia biology: one century of evolving concepts. Cell 179:292–311. https://doi.org/10.1016/j.cell.2019.08.053
2. Nayak D, Roth TL, McGavern DB (2014) Microglia development and function. Annu Rev Immunol 32:367–402. https://doi.org/10.1146/annurev-immunol-032713-120240
3. Hickman S, Izzy S, Sen P, Morsett L, El Khoury J (2018) Microglia in neurodegeneration. Nat Neurosci 21:1359–1369. https://doi.org/10.1038/s41593-018-0242-x
4. Stadelmann C, Timmler S, Barrantes-Freer A, Simons M (2019) Myelin in the central nervous system: structure, function, and pathology. Physiol Rev 99:1381–1431. https://doi.org/10.1152/physrev.00031.2018
5. Miller G (2005) Neuroscience. The dark side of glia. Science 308:778–781. https://doi.org/10.1126/science.308.5723.778
6. Lappe-Siefke C et al (2003) Disruption of Cnp1 uncouples oligodendroglial functions in axonal support and myelination. Nat Genet 33: 366–374. https://doi.org/10.1038/ng1095
7. Huang W et al (2020) Origins and proliferative states of human oligodendrocyte precursor cells. Cell 182:594–608.e511. https://doi.org/10.1016/j.cell.2020.06.027
8. Chen Y et al (2007) Isolation and culture of rat and mouse oligodendrocyte precursor cells. Nat Protoc 2:1044–1051. https://doi.org/10.1038/nprot.2007.149
9. Yazdankhah M et al (2021) BNIP3L-mediated mitophagy is required for mitochondrial remodeling during the differentiation of optic nerve oligodendrocytes. Autophagy 17:3140–3159. https://doi.org/10.1080/15548627.2020.1871204
10. Bohlen CJ, Bennett FC, Bennett ML (2019) Isolation and culture of microglia. Curr Protoc Immunol 125:e70. https://doi.org/10.1002/cpim.70
11. Goshi N, Morgan RK, Lein PJ, Seker E (2020) A primary neural cell culture model to study neuron, astrocyte, and microglia interactions in neuroinflammation. J Neuroinflammation 17: 155. https://doi.org/10.1186/s12974-020-01819-z

Chapter 14

Phenotypic and Functional Characterization of Memory $CD4^+$ and $CD8^+$ T Cells After Antigenic Stimulation

Carolina do Prado Servian, Letícia Carrijo Masson, and Simone Gonçalves Fonseca

Abstract

The encounter of T cells with the antigen through the interaction of T cell receptors with peptides and major histocompatibility complex (MHC) molecules on the surface of antigen-presenting cells (APCs) can generate effector response and memory T cells. Memory T cells developed following infections or vaccination may persist, leading to the generation of a specific immune response upon reexposure to the same pathogen through rapid clonal proliferation and activation of effector functions. T cell memory subsets can be identified based on the expression of several membrane markers such as CCR7, CD27, and CD45RA. Using fluorescent antibodies against these markers and a flow cytometer, it is possible to perform immunophenotyping via the analysis of cell surface expression of proteins by different subpopulations such as the subsets of naïve, effector, and memory T cells as well as via the analysis of functional markers that further characterize each sample. Intracellular cytokine staining allows for the evaluation of intracellular proteins expressed in T cells in response to antigenic stimulation. This chapter presents the phenotypic and functional characterization of memory T cells after antigenic stimulation, detailing the procedures for identifying intracellular and surface protein markers. Herein, we review and present a reproducible standardized protocol using antibodies for specific markers and applying flow cytometry.

Key words Antigen-specific, Memory T cells, $CD4^+$ T lymphocytes, $CD8^+$ T lymphocytes, Cell surface markers, Intracellular cytokine staining, Flow cytometry

1 Introduction

The differentiation of naïve T cells into the effector and memory subsets characterizes the immunity mediated by T cells. In response to antigenic stimulation, naïve T cells differentiate into the effector and memory subsets. Memory T cells can persist for years or even a lifetime and can rapidly react to previously encountered specific antigens with greater potency and efficacy [1–4]. The

Carolina do Prado Servian and Letícia Carrijo Masson contributed equally with all other contributors.

Vijay Kumar (ed.), *Immune Homeostasis: Methods and Protocols*, Methods in Molecular Biology, vol. 2782,
https://doi.org/10.1007/978-1-0716-3754-8_14,

Table 1
T cell heterogeneity among naïve and memory subsets

Surface marker	T_N	T_{SCM}	T_{CM}	T_{TM}	T_{EM}	T_{EMRA}	T_{RM}[a]
CCR7	+	+	+	−	−	−	−
CD45RA	+	+	−	−	+/−	+	−
CD27	+	+	+	+	−	−	−
CD95	−	+	+	+	+	+	+
CD69	−	−	−	−	−	−	+
CD103	−	−	−	−	−	−	+/−

[a]T_N naïve cells, T_{SCM} T memory stem cell, T_{CM} central memory T cells, T_{TM} transitional memory T cells, T_{EM} effector memory T cells, T_{EMRA} terminally differentiated effector memory T cells, T_{RM} tissue-resident memory T cells

differentiation of naïve T cells can result in distinct subsets of memory T cells known as memory stem cells (T_{SCM}), central memory (T_{CM}), transitional memory (T_{TM}), effector memory (T_{EM}), $CD45RA^+$ T_{EM} or terminally differentiated effector memory (T_{EMRA}), and also tissue-resident memory T (T_{RM}) cells that are found in the peripheral tissues. Memory and naïve T cells are classified based on the expression of different markers, mainly homing markers, chemokines, and cytokine receptors as well as on the release of distinct cytokines [1, 5–10]. The major markers of naïve and memory T cells are represented in Table 1.

Polyfunctional T cells are cells that have multiple simultaneous functions such as secretion of cytokines (IFN-γ, TNF-α, and IL-2) and chemokines, proliferation, and release of cytotoxic granules [11, 12]. It is suggested that these cells provide a more robust immunity since they can control some viral infections [11, 13]. The maintenance of memory cells can involve several factors; however, the mechanism is still not fully understood. Both naïve and memory T cells are dependent on signals emitted by cytokines to maintain survival for prolonged periods, although they are low-level signals for naïve cells [14–16]. For memory $CD8^+$T cells, IL-15 exerts this function, whereas IL-7 is responsible for the survival of memory $CD4^+$T cells [13–15]. These signals emitted by IL-15 and IL-7 are known as homeostatic signaling. IL-7 is mostly responsible for cell survival, while IL-15 is more involved with homeostatic cell proliferation [17]. Additional cytokines, such as IL-2, IL-10, and IL-12, interferons and others can also promote this state of homeostasis [14–16].

Flow cytometry is a technique widely used to evaluate multiple cell features and to identify several cell population subsets. It is an assay based on fluorescence emission, captured by photon detectors, after fluorochrome excitation by laser beams of different wavelengths. These fluorescent molecules are connected to antibodies that have a specific antigenic target on the evaluated cells, thereby promoting the identification of a particular cell population [18].

The identification of memory T cell subsets can be performed by flow cytometry immunophenotyping. An analysis of the association of lymphocyte function with T cell phenotype can help with the identification of these subsets. Cell surface staining is an assay used to detect specific cell populations by flow cytometry, and it is based on the staining of markers or antigens, present on the cell surface, by antibodies bonded to fluorochromes. Intracellular cytokine staining (ICS) is also an assay that enables the identification of different cell subsets since it provides a simultaneous evaluation of markers of cell differentiation and proliferation as well as the detection of cytokines or chemokines for function assessment [19]. Combined, these two protocols can provide more detailed identification of cell subsets. However, given that a membrane permeabilization step is necessary to perform ICS, caution must be taken when performing these two techniques together, in order not to make the detection of surface antigens unfeasible due to permeabilization and partial loss of signal [20].

In this chapter, we focus on the phenotypic and functional identification of T cell subsets after antigenic stimulation by flow cytometry. We describe antigen stimulation of human peripheral blood mononuclear cells (PBMC) as well as the phenotypic and functional characterization of naïve and memory $CD4^+$ and $CD8^+$ T cell subsets using the multiparameter flow cytometry protocol. Here, we describe two panels of antibodies that can be used to perform cell surface and intracellular cytokine staining assays.

2 Materials

2.1 Reagents

1. Sample material (*see* **Note 1**).
2. Culture medium [commercial Roswell Park Memorial Institute (RPMI) 1640 medium supplemented with 20 mM HEPES, 5% fetal bovine serum (FBS) heat-inactivated, 2 mM L-glutamine, 100 U ml^{-1} penicillin, 100 U mL^{-1} streptomycin].
3. Phosphate-buffered saline (1X PBS), pH 7.4.
4. Fluorescence-activated cell sorter (FACS) buffer (1X PBS in 2% FBS).
5. Kit BD Cytofix/Cytoperm (BD Biosciences, San Diego, USA).
6. Golgi Plug – Brefeldin A (BD Biosciences, San Diego, USA) (*see* **Note 2**).
7. Negative control for peptide stimulation: Dimethyl sulfoxide (DMSO, Sigma) since peptides are diluted in DMSO.

Table 2
Antibody panel details: monoclonal antibodies used in this protocol

Antibodies	Fluorochrome	Source	Ab Clone	Catalog	Volume
Antibodies for surface staining					
CD3	V450	BD horizon	UCHT1	560365	1 μL
CD4	PerCPCy5.5	BD Pharmingen	RPA-T4	560650	1 μL
CD8	PE	BD biosciences	HIT8a	555635	1 μL
CD45RA	APC-H7	BD Pharmingen	HI100	560674	2 μL
CD27	V500	BD horizon	M-T271	561222	3 μL
CCR7	PE-Cy7	BD Pharmingen	3D12	560922	3 μL
Antibodies for intracellular staining					
IL-2	APC	BD FastImmune	5344.111	341116	3 μL
IFN-γ	FITC	BD Pharmingen	B27	554700	1 μL

APC-Cy7 allophycocyanin-cyanine H7 tandem, *PerCPCy5.5* peridinin chlorophyll protein-cyanine 5.5, *PE* phycoerythrin, *APC-H7* allophycocyanin H7, *V500* BD horizon dye V500, *PE-Cy7* phycoerythrin-cyanine 7 tandem, *APC* allophycocyanin, *FITC* fluorescein isothiocyanate. Each antibody conjugated was added to the cells in a final volume of 25 μL (each antibody + FACS buffer)

8. Antigenic stimulation: peptide or peptide pool, such as HIV peptide pool [21], diluted in DMSO, at concentration previously titrated. Positive control for stimulation: *Staphylococcus enterotoxin* B (SEB) (*see* **Note 3**)
9. PermWash (BD Biosciences, San Diego, USA) (*see* **Note 4**).
10. 37% formaldehyde solution [2% (w/v) formaldehyde in 1X PBS] (Sigma-Aldrich, St. Louis, USA) (*see* **Note 5**)
11. Polypropylene tube (Falcon, cat. 352063, Corning, USA) (*see* **Note 6**).
12. Polystyrene FACS tube (Falcon, Corning, USA).
13. 15 mL Falcon tubes.
14. 1.5 mL Eppendorf tubes.
15. Multichannel and single channel pipettes.
16. Fluorochrome-conjugated monoclonal antibodies (*see* Table 2).

2.2 Equipment

1. Refrigerated centrifuge.
2. Vortex.
3. 5% CO_2 incubator at 37 °C.

4. Flow cytometer: Fluorescence-activated cell sorter (FACS). The BD FACSCanto™ II equipped with FACSDiva software (BD Biosciences, San Diego, USA) was used in the experiment displayed in this chapter (*see* **Note 7**).
5. FlowJo™ v10 analysis software (BD Biosciences, San Diego, USA).

3 Methods

3.1 Antigenic Stimulation of Peripheral Blood Mononuclear Cells (PBMC)

The cell separation and antigenic stimulation should be done in the biological safety cabinet.

1. Thaw the PBMCs in a 15 mL falcon tube with 10 mL of RPMI supplemented with antibiotics and 10% fetal bovine serum (FBS); wash three times with complete RPMI by centrifugation at 800 × *g* for 8 min; count the cells using trypan blue in hemocytometer chamber for checking viability and incubate the cells at a concentration of 2×10^6 cells/mL of RPMI supplemented for at least 2 h at 37 °C in a 5% CO_2 prior to stimulation. After count the cells using trypan blue in hemocytometer chamber for checking viability.
2. Centrifuge the cells at 800× *g* for 8 min and resuspend cell pellets in RPMI (supplemented with antibiotics and 5% FBS) after the new count; add 2×10^6 cells (*see* **Note 8**) to each of the polypropylene tubes with caps – the final volume should be 1 mL Cell activation can also be done in U-botton 96 well plate 2×10^6 cells/200 ul of medium.
3. Stimulate the cells (Fig. 1) with a volume of the pool of peptides according to the previous concentration titrated (*see* **Note 9**), such as the HIV peptide pool [21].
4. As a positive control, stimulate the cells with 5 ug/mL of staphylococcal enterotoxin B from *Staphylococcus aureus* (SEB, Sigma) (2.5 μL of SEB, which is at 2 mg/mL in sterile

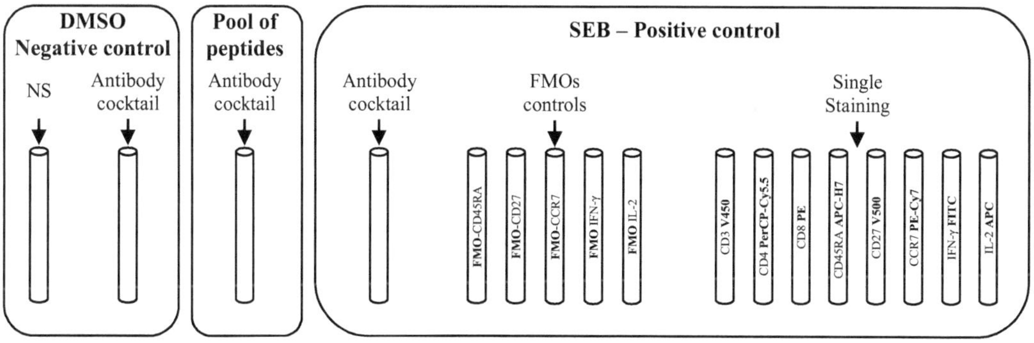

Fig. 1 Design of basic protocol for flow cytometry staining

PBS). All tubes for fluorescence minus one (FMO) controls should be stimulated with SEB.

5. Non-stimulated (NS) cells should be incubated with the same volume of dimethyl sulfoxide (DMSO, Sigma) used for diluting the peptide pool for NS.
6. Leave the cells for 1 h in a 5% CO_2 incubator at 37 °C.
7. After 1 h of stimulation, add 1 μg/mL of Golgi Plug – Brefeldin A (1 μL) to all tubes.
8. Incubate the tubes for 4 or 12 h at 37 °C in a 5% CO_2 incubator (*see* **Note 10**).
9. Remove 10 μL from each tube to perform cell viability using trypan blue (*see* **Note 11**).

3.2 Cell Surface Staining

After stimulation with a pool of peptides, SEB, or DMSO, the PBMCs must be labeled with a panel of surface antibodies (*see* Table 2) to evaluate the T cell subsets.

The staining should be done in the dark. Switch off the light from the biological safety cabinet.

1. For surface staining, transfer the cells to polystyrene tubes, add 1 mL of FACS buffer, and centrifuge at 800× *g* for 5 min. Alternatively, the staining can be done in 96-well V bottom plate.
2. Discard supernatant; add 1 mL of FACS buffer (PBS + 2% FBS) to each tube.
3. Centrifuge at 800× *g* for 5 min.
4. Discard the supernatant and gently invert the tubes on paper towels (*see* **Note 12**).
5. Repeat **steps 2–4** (1×).
6. For surface staining, add to the cells 25 μL of surface antibody cocktail containing the fluorochrome-labeled antibodies at a standardized concentration (*see* Table 2) as predetermined by titration experiments. To prepare the surface antibody cocktail, always calculate the number of samples that will be stained and add an extra volume of each antibody. For example, if using 20 samples, add enough antibodies to stain 22 samples. The surface antibody cocktail should be prepared in Eppendorf tubes considering the volume of each surface antibody previously titrated anti-CD3 (V450), CD4 (PerCP-Cy5.5), CD8 (PE), CD45RA (APC-H7), CD27 (V500), and CCR7 (PE-Cy7) and be completed to a final volume of 25 μL/tube with FACS buffer. Pipette 25 μL of the surface antibody cocktail solution with anti-human monoclonal antibodies conjugated with fluorochromes and add to the tubes containing the cells stimulated with SEB, the pool of peptide, and

Table 3
Fluorescence minus one (FMO) controls used in this protocol

Fluorochrome	1	2	3	4	5	6
APC-H7	CD45RA	**FMO APC-H7**	CD45RA	CD45RA	CD45RA	CD45RA
V500	CD27	CD27	**FMO V500**	CD27	CD27	CD27
PE-Cy7	CCR7	CCR7	CCR7	**FMO PE-Cy7**	CCR7	CCR7
APC	IL-2	IL-2	IL-2	IL-2	**FMO APC**	IL-2
FITC	IFN-γ	IFN-γ	IFN-γ	IFN-γ	IFN-γ	**FMO FITC**

The FMO control for each fluorochrome is indicated in **bold**. For those tubes, the samples will be stained with every fluorochrome except the fluorochrome indicated in **bold**. *FMO* fluorescence minus one

DMSO, and lightly vortex to resuspend. The tubes should be named as:

- Tube for NS – to the unstained and unstimulated cells, add 25 of FACS buffer (*see* **Note 13**).
- Tubes for each single staining – for compensation, use unstimulated cells and a single-color control for each antibody concentration used in the cocktail: anti-CD3 (V450), anti-CD4 (PerCP-Cy5.5), anti-CD8 (PE), anti-CD45RA (APC-H7), anti-CD27 (V500), and anti-CCR7 (PE-Cy7).
- Tube for each FMO control (*see* Table 3) – label the sample with all the antibodies and dyes except for one (*see* **Note 14**) (Fig. 1).

7. Incubate the tubes for 30 min at 4 °C in the dark (wrap aluminum foil around the rack with the tubes).
8. Add 1 mL of FACS buffer to each tube.
9. Centrifuge at 800× *g* for 5 min.
10. Discard the supernatant and gently invert the tubes onto paper towels.
11. Repeat **steps 8–10** (1×).
12. Proceed with the protocol for intracellular labeling (as per the next step of the protocol).

3.3 Intracellular Cytokine Staining

1. Add 100 μL of Cytofix/Cytoperm (dark tube – already diluted for use) 1× to all tubes.
2. Vortex all tubes at low speed.
3. Incubate for 20 min at 4 °C in the dark.

4. After incubation, add 1 mL of 1× PermWash to each tube and centrifuge for 5 min at 800× *g* (PermWash also on NS – unstained and unstimulated cells, and unstimulated tubes).
5. Discard the supernatants and invert the tubes onto paper towels.
6. Repeat **steps 4** and **5**.
7. The supernatant is discarded and the cytokine antibody cocktail with anti-IFN-γ (FITC) and anti-IL-2 (APC) is added (*see* **Note 15**) at a standardized concentration as predetermined (Table 2). To prepare the intracellular antibody cocktail in Eppendorf tubes, always calculate how many samples will be stained and add an extra volume of each antibody. For example, if using 20 samples, add enough antibodies to stain 22 samples. Pipette the solution of antihuman monoclonal antibodies conjugated with fluorochromes and add it to the cells in a final volume of 25 μL (each intracellular antibody cocktail volume and should be completed to 25 μL/tube with PermWash 1×) in each tube (SEB, DMSO, or pool of peptides) and lightly vortex to resuspend (Fig. 1). To a tube with non-stained cells should be added 25 μL of PermWash.
   - Tube unstimulated (DMSO), tube with a pool of peptides, and tube with SEB – staining procedure with the cytokine antibody cocktail: anti-IFN-γ (FITC) and IL-2 (APC).
   - Tube single staining – for compensation, use unstimulated cells and single-color control for each antibody concentration used in the cocktail: anti-IFN-γ (FITC) and IL-2 (APC).
   - Tube FMO controls (*see* Table 3) – label the sample with all the antibodies and dyes except for one (*see* **Note 14**) (Fig. 1).
8. Incubate for 30 min at 4 °C in the dark.
9. After incubation, add 1 mL of 1× PermWash to all tubes.
10. Centrifuge for 5 min at 800× *g*.
11. Discard the supernatant and invert tubes onto paper towels.
12. Repeat washing.
13. Add 200 μL of 2% formaldehyde (diluted in 1 × PBS) solution to each tube.
14. If the cytometer reading is not taken on the same day, store the tubes at 4 °C in the dark for reading on the cytometer the next day. Acquire cells at an event rate equal to or higher than 500,000 events on a BD FACSCanto II flow cytometer (BD Biosciences, San Diego, USA) and record data. The description of the eight-color panel to identify the subsets is described in Table 4.

Table 4
Flow cytometer configuration

Excitation laser line (nm)	Laser power (mW)	Fluorescence detection (nm)	Fluorochrome
488	20	515–545	FITC
		564–606 670–735 750–810	PE PerCPCy5.5 PE-Cy7
633	17	650–670 750–810	APC APC-H7
405	30	425–475 502–535	V450/Pacific Blue V500/AmCyan

The panel of antibodies used in the protocol described in this chapter was optimized for a BD FACSCanto II with the listed optical elements

3.4 Data Analysis

3.4.1 Creating Gates

1. After data acquisition, the results should be analyzed on a specific cytometry software such as FlowJo (BD Biosciences, San Diego, USA). Usually, forward scatter area (FSC-A) and side scatter (SSC-A) are parameters used in the initial gates, since they delimit the cell size and granularity, respectively [13], [17]. FSC-A is typically used on the X-axis and SSC-A on the Y-axis. Subsequently, a gate should be created to initially select the lymphocyte population (*see* Table 4).
2. Inside the lymphocyte gate, to remove cell doublets and eventual debris, the singlet gate should be applied between FSC-A and forward light scatter height (FSC-H).
3. Next, to identify the T cell population, $CD3^+$ cells should be selected.
4. Inside the population, $CD3^+$, $CD4^+$, and $CD8^+$-T cells should be differentiated based on the expression of these markers.
5. The next gate should be performed based on the expression of CD45RA as $CD45RA^+$ or $CD45RA^-$.
6. Inside $CD45RA^+$ or $CD45RA^-$, the subsets should be gated according to expression of CCR7 and CD27 to identify T cell memory subsets, accordingly.

   T_N: $CD45RA^+$ $CCR7^+$ $CD27^+$.

   T_{EMRA}: $CD45RA^+$ $CCR7^-$ $CD27^-$.

   T_{CM}: $CD45RA^-$ $CCR7^+$ $CD27^+$.

   T_{EM}: $CD45RA^-$ $CCR7^-$ $CD27^-$.
7. Each of these subsets of $CD4^+$ and $CD8^+$ T cells should be analyzed for the expression of cytokines such as IL-2 and

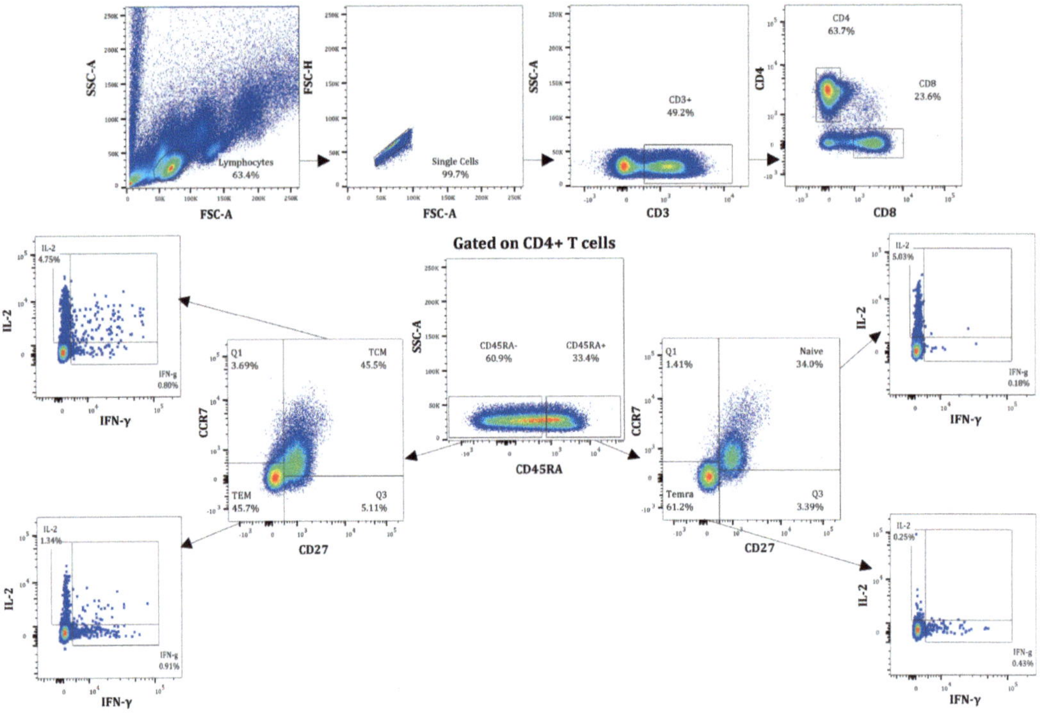

Fig. 2 Gating strategy for phenotypic and functional characterization of T cell memory subsets after antigenic stimulation. After the sequence of gates inside the subsets of CD4$^+$ and CD8$^+$ T cells, show the frequency of the cytokines after antigenic stimulation in each memory subset. CD45RA, CCR7, and CD27 help in identifying T_N, T_{EMRA}, T_{CM} and T_{EM} in the peripheral blood of healthy individuals. For representative examples of cytokine production evaluated by intracellular staining, the expression of IFN-γ and IL-2 was evaluated in each subsets. The gates listed above indicate the gating hierarchy from one gate to the next. The combination of the forward scatter area (FSC-A) with the side scatter (SSC-A) is utilized as a gate for lymphocytes. Next, FSC-A and forward scatter height (FSC-H) are utilized to exclude doublets. In this figure, only memory subsets of CD4$^+$ T cells are represented. For CD8$^+$ T cells, the gate strategy should be similar. In this figure PBMC were stimulated with staphylococcal enterotoxin B (SEB). Abbreviations: T_N (naive), T_{EMRA} (terminally differentiated memory cells), T_{CM} (central memory) and T_{EM} (effector memory)

IFN-γ.(Fig. 2). An alternative method of analysis involves gating on cytokines produced by CD4$^+$ or CD8$^+$ T cells, followed by the analysis of memory T cell subsets (Fig. 3). This latter approach allows to define the memory profile of antigen-specific T cells. Representative flow cytometry plots showing cytokine expression by CD4$^+$ and CD8$^+$ T cells after specific antigenic stimulation (Fig. 4).

8. The cytokine production under antigen stimulation should be subtracted from cells cultured in the presence of DMSO (control). A positive response is considered when the response is at least 2 times greater than that obtained with non-stimulated cells (DMSO) For statistical analysis, use the software of your preference.

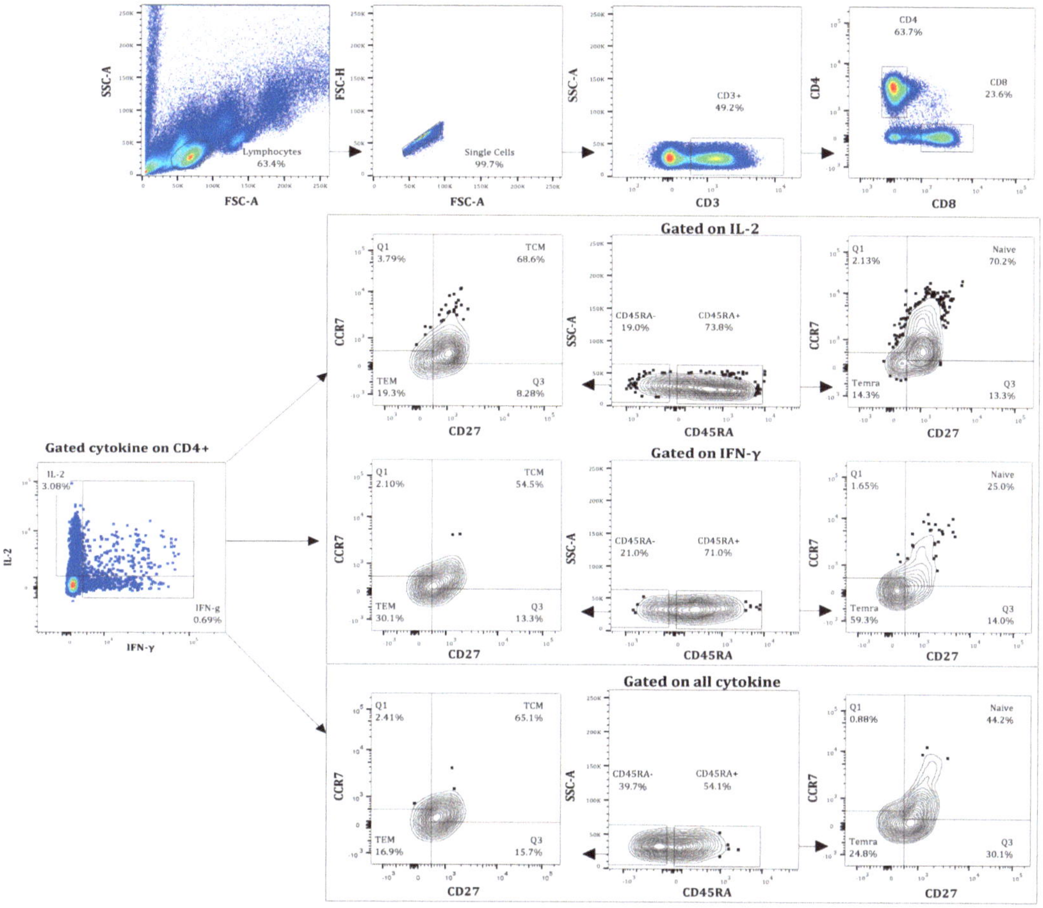

Fig. 3 Gating strategy for characterization of memory T cell subsets that produce cytokines in response to stimuli. The combination of forward scatter area (FSC-A) with side scatter area (SSC-A) was utilized as a gate for lymphocytes. Then, FSC-A and forward scatter height (FSC-H) were utilized to exclude doublets. $CD3^+$ T cells were gated, and within this population, $CD4^+$ and $CD8^+$ T cells were gated. Within the $CD4^+$ or $CD8^+$ T cell populations, cells producing cytokines were gated. Gating for IL-2, IFN-γ producing cells, or cells producing all cytokines using Boolean gating was performed. Within the cytokine-producing population, gates were set for subsets of memory T cells and naÿve cells using CD45RA, CCR7, and CD27. PBMCs were stimulated with Staphylococcal enterotoxin B (SEB) at 5 ug/ml. PBMCs non-stimulated were used as a control

4 Notes

1. In this protocol, we used cryopreserved PBMC isolated through Ficoll gradient separation. PBMC thawing was performed in a water bath at 37 °C until there was a small ice pellet remaining. Freshly isolated PBMC can also be used to perform this protocol, such as PBMC obtained through heparinized whole blood.
2. In the absence of Brefeldin A, GolgiPlug and GolgiStop (BD Biosciences) or monensin can be used.

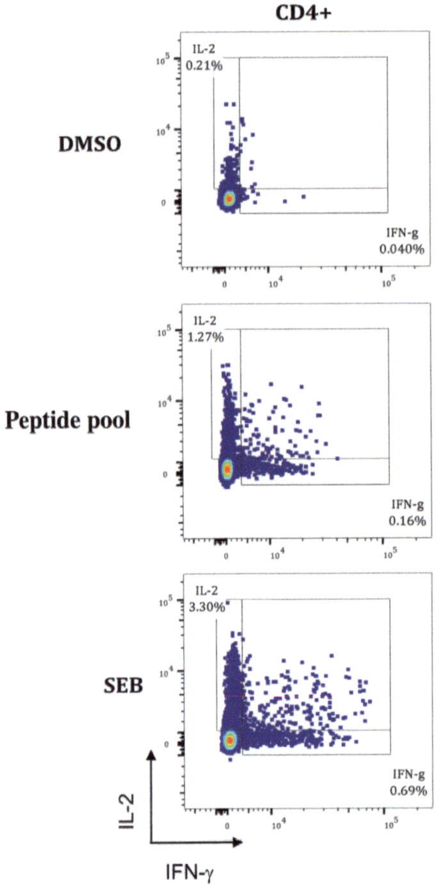

Fig. 4 Representative flow cytometry plots showing cytokine expression by $CD4^+$ and $CD8^+$ T cells after specific antigenic stimulation. Peripheral blood mononuclear cells (PBMCs) were stimulated with 1ug/ml of SARS-CoV peptide pool, 5 ug/ml of Staphylococcal enterotoxin B (SEB) as a positive control, or non-stimulated (in presence of DMSO) as a negative control. After a 12-hour incubation period, cells were stained with antibodies against anti-CD3 V450, anti-CD4 PerCPCy5.5, and anti-CD8 PE as surface markers, and anti-IL-2 APC and IFN-Y FITC following the protocol for intracellular staining. Gate strategy for $CD4^+$ and $CD8^+$ T cells was performed as shown in Fig. 2. Cytokine-producing cells are shown within the CD4 and CD8 gates. The percentages of cells producing IL-2 and IFN-γ are shown

3. Anti-CD3/CD28, Ionomycin, or phytohemagglutinin (PHA) at 1 μg/mL can be used as activators for positive control.
4. Saponin, at a concentration of 0.1%, is another solution that can be used as a permeabilization buffer for cells.
5. Cells can be fixed by the addition of 1% paraformaldehyde diluted in PBS.
6. Other options like a 96-well U bottom plate (200 μL/well) can be used.
7. Other flow cytometers available for detecting other panels of fluorochromes can be used (eight colors at least).

8. Cell count should be at a concentration of 2×10^6 cells/mL. Nevertheless, if not possible, the total cell count should be at least 1×10^6 cells/mL.
9. Other concentrations of the peptide pool can be used as previously characterized and standardized. The final concentration of DMSO in the tube should be no more than 1/500.
10. For the simulation time, cells can be incubated for any period that is convenient for the analysis, usually a minimum of 5–6 h to allow for adequate stimulation.
11. LIVE/DEAD Aqua Vivid should preferably be used to evaluate the cell viability. In our case, it was not used because it has a fluorescence emission that is similar to the other fluorochromes used in the cocktail. Other cell viability dyes, from other companies, can be used, such as amine reactive; fixable amine-binding viability dyes: for example, Zombie (Biolegend, San Diego, CA) and fixable viability (BD Biosciences).
12. If the cells being used in the protocol present a risk of contamination, the culture supernatant should be discarded in a biohazard waste container under a laminar flow hood.
13. For compensation, reference beads such as BD™ CompBead can alternatively be used.
14. The FMO strategy is used in multicolor flow cytometric analysis to set boundaries between positive fluorescence and negative fluorescence as well as to set up the positive and negative populations.
15. Other cytokines can be used in the intracellular cytokine staining assay, such as TNF-α, IL-4, IL-21, IL-17A, IL-10, IL-13 and other cytokines or other molecules such as granzyme B and perforin.

Acknowledgements

We are thankful to Dr. Catherine Riou from the Institute of Infectious Disease and Molecular Medicine at the University of Cape Town, South Africa, for her invaluable assistance in teaching us the FACS analysis presented in Fig. 3.

Funding

This work was supported by the Fundação de Amparo à Pesquisa do Estado de Goiás (FAPEG), grant ID 202010267000284. Carolina do Prado Servian and Letícia Carrijo Masson received scholarships from Coordenação de Aperfeiçoamento de Pessoal de Nível Superior (CAPES). Simone GonÓalves Fonseca is a research fellow of the Conselho Nacional de Desenvolvimento Científico e Tecnológico (CNPq).

References

1. Künzli M, Masopust D (2023) CD4+ T cell memory. Nat Immunol 24:903–914. https://doi.org/10.1038/s41590-023-01510-4
2. Tanel A, Fonseca SG, Yassine-Diab B et al (2009) Cellular and molecular mechanisms of memory T-cell survival. Expert Rev Vaccines 8:312. https://doi.org/10.1586/14760584.8.3.299
3. Dutton RW, Bradley LM, Swain SL (1998) T cell memory. Annu Rev Immunol 16:201–223. https://doi.org/10.1146/ANNUREV.IMMUNOL.16.1.201
4. Derksen LY, Tesselaar K, Borghans JAM (2023) Memories that last: dynamics of memory T cells throughout the body. Immunol Rev 316:38–51. https://doi.org/10.1111/imr.13211
5. Sallusto F, Geginat J, Lanzavecchia A (2004) Central memory and effector memory T cell subsets: function, generation, and maintenance. Annu Rev Immunol 22:745–763. https://doi.org/10.1146/ANNUREV.IMMUNOL.22.012703.104702
6. Sallusto F, Lenig D, Förster R et al (1999) Two subsets of memory T lymphocytes with distinct homing potentials and effector functions. Nature 401:708–712. https://doi.org/10.1038/44385
7. Buggert M, Price DA, Mackay LK et al (2023) Human circulating and tissue-resident memory CD8+ T cells. Nat Immunol 24:1076–1086. https://doi.org/10.1038/s41590-023-01538-6
8. Farber DL, Yudanin NA, Restifo NP (2013) Human memory T cells: generation, compartmentalization, and homeostasis. Nat Rev Immunol 14:24–35. https://doi.org/10.1038/nri3567
9. Gattinoni L, Lugli E, Ji Y et al (2011) A human memory T cell subset with stem cell-like properties. Nat Med 17:1290–1297. https://doi.org/10.1038/NM.2446
10. Appay V, Van Lier RAW, Sallusto F, Roederer M (2008) Phenotype and function of human T lymphocyte subsets: consensus and issues. Cytometry A 73:975–983. https://doi.org/10.1002/CYTO.A.20643
11. Lam JKP, Hui KF, Ning RJ et al (2018) Emergence of CD4+ and CD8+ polyfunctional T cell responses against immunodominant lytic and latent EBV antigens in children with primary EBV infection. Front Microbiol 9:1–13. https://doi.org/10.3389/FMICB.2018.00416/BIBTEX
12. Imai N, Tawara I, Yamane M et al (2020) CD4+ T cells support polyfunctionality of cytotoxic CD8+ T cells with memory potential in immunological control of the tumor. Cancer Sci 111:1958–1968. https://doi.org/10.1111/CAS.14420
13. Boyd A, Almeida JR, Darrah PA et al (2015) Pathogen-specific T cell Polyfunctionality is a correlate of T cell efficacy and immune protection. PLoS One 10:1–18. https://doi.org/10.1371/JOURNAL.PONE.0128714
14. Kawabe T, Yi J, Sprent J (2021) Homeostasis of naive and memory T lymphocytes. Cold Spring Harb Perspect Biol 13(9). https://doi.org/10.1101/cshperspect.a037879
15. Boyman O, Purton JF, Surh CD, Sprent J (2007) Cytokines and T-cell homeostasis. Curr Opin Immunol 19:320–326. https://doi.org/10.1016/J.COI.2007.04.015
16. Radbruch A, McGrath MA, Siracusa F et al (2021) Homeostasis and durability of T-cell memory-the resting and the restless T-cell memory. Cold Spring Harb Perspect Biol 13(7). https://doi.org/10.1101/cshperspect.a038083
17. Surh CD, Sprent J (2008) Homeostasis of naive and memory T cells. Immunity 29:848–862. https://doi.org/10.1016/J.IMMUNI.2008.11.002
18. McCoy JP Jr (2019) Immunophenotyping, 1st edn. Springer, New York
19. Smith SG, Smits K, Joosten SA et al (2015) Intracellular cytokine staining and flow cytometry: considerations for application in clinical trials of novel tuberculosis vaccines. PLoS One 10:1–20. https://doi.org/10.1371/JOURNAL.PONE.0138042
20. Krutzik PO, Clutter MR, Nolan GP (2005) Coordinate analysis of murine immune cell surface markers and intracellular phosphoproteins by flow cytometry. J Immunol 175:2357–2365. https://doi.org/10.4049/JIMMUNOL.175.4.2357
21. Al-Kolla R, Grifoni A, Crotty S et al (2022) Design and validation of HIV peptide pools for detection of HIV-specific CD4+ and CD8+ T cells. PLoS One 17:1–12. https://doi.org/10.1371/JOURNAL.PONE.0268370

Chapter 15

CRISPR-Cas9-Induced Gene Editing in Primary Human Monocytes

Xiaofeng Liao and Liwu Li

Abstract

Monocytes play important and diverse roles in both homeostatic and inflammatory immune responses. The CRISPR-Cas9 system in lentiviral vectors has been widely used to manipulate specific genes of immortal monocyte cell lines to study monocyte functions. However, human primary monocytes are refractory to this method with low gene knockout (KO) efficiency. In this chapter, we developed an in vitro gene-editing procedure for primary human monocytes with a consistent and high-gene KO efficiency via a ribonucleoprotein (RNP) complex consisting of Cas9 protein and single-guide RNA (sgRNA). This method can be adapted to study the functions of targeted signaling molecules involved in modulating monocyte polarization in primary human monocytes.

Key words CRISPR-Cas9, Monocyte, Memory dynamics, Culture system, Transfection, Gene editing

1 Introduction

There are emerging interests in studying the memory dynamics of primary innate immune cells such as monocytes, given their vital contributions during the pathogenesis of acute as well as chronic diseases [1, 2]. In order to precisely define underlying molecular mechanisms responsible for the generation of innate monocyte memory dynamics, genetic manipulations of key signaling molecules are necessary. Although it is relatively easy to generate the deletion of targeted genes in the murine system either in vitro or in vivo, it has been a challenge with primary human monocytes due to the limited transfection and targeting efficiency.

Recently developed CRISPR-Cas9 system via lentiviral transduction offers significant potential in performing targeted gene-editing in both primary and immortalized cells [3]. However, its targeting efficacy with primary human monocytes is less tested. To refine this system, we developed an in vitro gene-editing procedure

Vijay Kumar (ed.), *Immune Homeostasis: Methods and Protocols*, Methods in Molecular Biology, vol. 2782,
https://doi.org/10.1007/978-1-0716-3754-8_15,

suitable for primary human monocytes with a consistent and high targeting efficiency via a ribonucleoprotein (RNP) complex consisting of Cas9 protein and single-guide RNA (sgRNA).

2 Materials

Primary human blood samples: human unpurified buffy coat from health donors (research blood components).

Monocyte isolation supplies: 1×PBS, Ficol-plus, 1×PBS + 2% FBS, magnetic-activated cell sorting (MACS) buffer (1×PBS, 0.5% BSA, 2 mM EDTA), human pan-monocyte isolation kit (Miltenyi Biotec, Cat. 130-096-537).

Transfection: IDTE buffer, Cas9 protein, sgRNAs and Alt-R Cas9 electroporation enhancer (IDT).

P3 buffer with supplementl, 16-well strip, 4D-x unit Lonza nucleofector (Lonza).

Culture medium: RPMI1640, Glutamax, FBS, PenStrep (PS) antibiotics, recombinant human M-CSF (rhMCS-F) (R&D, Cat.216-MC).

Stimulation: LPS (0111:B4, Sigma).

3 Methods

3.1 Primary Human Monocyte Isolation

1. 15–20 mL buffy coat of human blood is 1:1 diluted at room temperature (RT) with 1×PBS and layered on 15 mL RT Ficol-Plus in a 50 mL SepMate (stem cell) tube.
2. Centrifuge at 1200 g for 10 min under RT with brake on.
3. Pour off the top layer containing peripheral blood mononuclear cells (PBMC) into a new 50 mL tube.
4. Wash cells twice with 1×PBS + 2% FBS, with centrifugation at 300 g × 10 min under RT each time to collect the cells.
5. Resuspend cells in 10 mL of ice-cold magnetic activated cell sorting (MACS) buffer and count cell number.
6. Centrifuge 300 g × 5 min under 4 °C and proceed to the monocyte purification step following the manufacturer's instruction.
7. Count purified monocytes in a collection tube and aliquot 2×10^4 to 1×10^5 cells for purity verification by flow cytometry staining of CD14-APC (Biolegend, #301807) and CD16-PE-Cy7 (Biolegend, #302015) (Fig. 1).

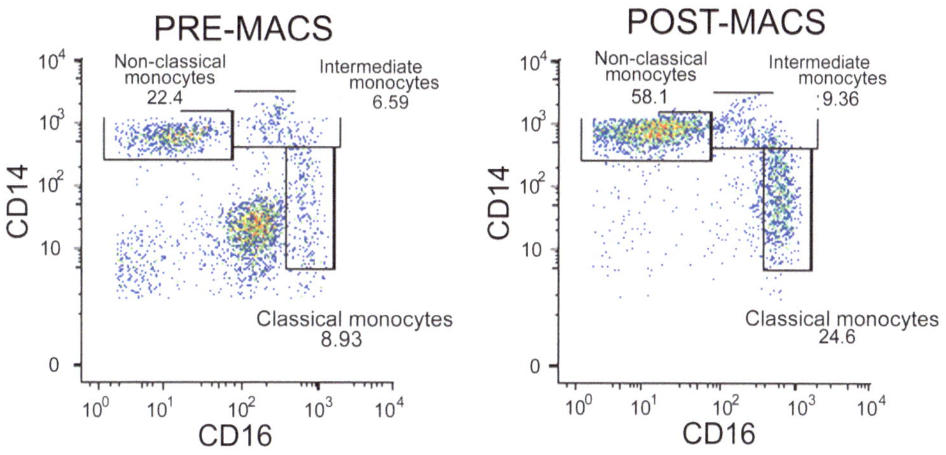

Fig. 1 Primary human monocyte isolation Flow cytometry analyses in detecting the percentages of monocyte subtypes based on CD14 and CD16 levels before (left panel, pre-MACS) and after (right panel, post-MACS) MACS-based purification

3.2 Gene Knockout (KO) of Primary Human Monocytes by Nucleofection

1. Prepare monocyte culture medium: RPMI1640 containing 10% FBS, 1% PS, 1% Glutamax, 10 ng/mL rhMCS-F.
2. Pre-warm 1 mL of monocyte medium in a 24-well plate in a 37 °C cell culture incubator.
3. Prepare Cas9: sgRNA RNP:
   (a) Dissolve sgRNA in sterile IDTE buffer to 100 μM concentration.
   (b) Mix 2.1 μL RT 1×PBMC, 1.2 μL sgRNA and 1.7 μL Cas9 protein in a sterile PCR tube.
   (c) Incubate under RT for 15–20 min to form RNP. In the meantime, prepare other steps.
4. To prepare P3 buffer, mix 16.4 μL P3 main buffer and 3.6 μL supplement 1 and keep at RT.
5. Turn on the Lonza 4D nucleofector and select reaction wells and program EA-100.
6. Spin down 1×10^6 monocytes 300 g × 5 min at RT.
7. Aspirate medium and leave about 50 μL to spin down for 1 min without resuspension.
8. Use 200 μL pipet tip to take the remaining medium out completely without disrupting the cell pellet.
9. Resuspend cells in 20 μL P3 buffer and mix with 5 μL RNP complex and 1 μL electroporation enhancer.
10. Use a thin tip to transfer the cell suspension completely into the nucleocuvette vessel without introducing bubbles.
11. Start electroporation immediately and remove pre-warmed culture medium.

12. Add 100 μL pre-warmed culture medium into cells immediately post-electroporation.
13. Transfer cell suspension into a 24-well plate containing pre-warmed culture medium.
14. Keep in culture for 3 days before confirmation of KO efficiency.

3.3 Functional Study of Primary Human Monocytes Post-nucleofection-Induced KO

1. After 3 days of culture, 2×10^5 monocytes are used to test gene KO efficiency at the protein level by western blot (Fig. 2).
2. Seed 5×10^4 monocytes/100 μL/well of non-tissue culture treated 96-well flat bottom plate.
3. Treat with or without 100 ng/mL LPS stimulation overnight.
4. Harvest cells for flow cytometry staining to detect phenotype changes (Fig. 3).

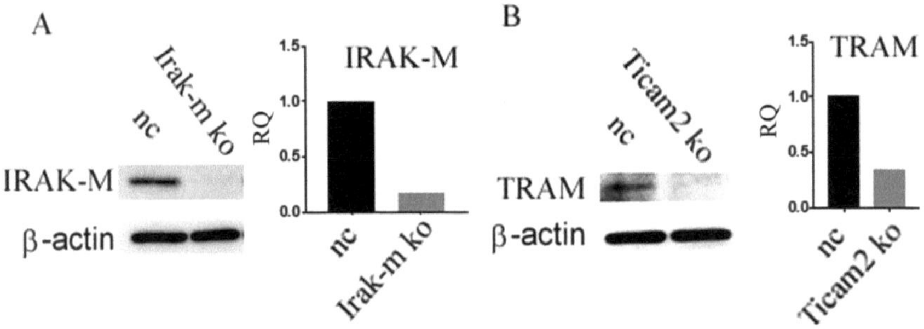

Fig. 2 Gene knockout (KO) of primary human monocytes by nucleofection
Western blot analyses of IRAK-M (**a**) and TRAM (TICAM2) (**b**) in primary human monocytes targeted by Cas9-sgRNA RNP electroporation. Beta-actin serves as an internal control for the calculation of relative protein quantity (RQ) in the right panel of each figure

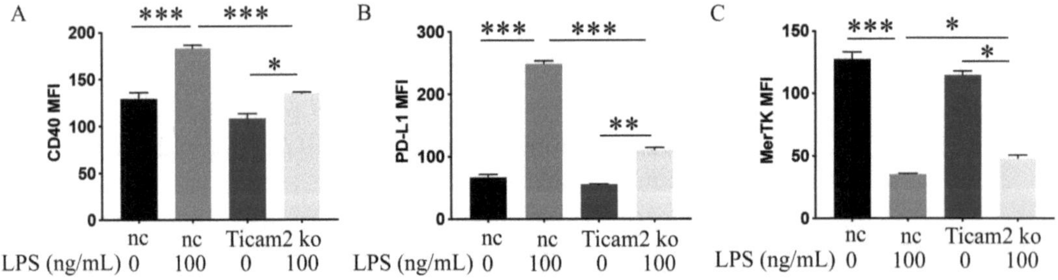

Fig. 3 Functional study of primary human monocytes post-nucleofection-induced KO
Flow cytometry measurement of CD40 (**a**), PD-L1 (**b**) and MerTK (**c**) levels on non-targeting control (nc) or Ticam2 KO monocytes with or without 100 ng/mL LPS stimulation overnight.

4 Notes

1. Aspirate medium completely before re-suspending cells in electroporation buffer, otherwise remaining trace medium will reduce KO efficiency.
2. Do not let cells stay in the electroporation buffer for a long time, which will reduce cell viability post-electroporation.
3. Warm-up recovery medium to 37 °C in a 5% CO_2 incubator before the nucleofection procedure to increase cell viability post-electroporation.
4. Transfer cells in a 16-well strip without any bubbles using thin tips that can touch the bottom of the well and completely cover the bottom of the well. Otherwise KO efficiency will be reduced.
5. Add warmed recovery medium immediately after electroporation with only one-time pipetting because cells are fragile right post-electroporation.

References

1. Karlmark KR, Tacke F, Dunay IR (2012) Monocytes in health and disease – minireview. Eur J Microbiol Immunol 2:97–102
2. Kapellos TS et al (2019) Human monocyte subsets and phenotypes in major chronic inflammatory diseases. Front Immunol 10:2035
3. Cong L, Zhang F (2015) Genome engineering using CRISPR-Cas9 system. Methods Mol Biol 1239:197–217

Chapter 16

Employing CRISPR-Cas9 to Enhance T Cell Effector Function

Julian J. Freen-van Heeren

Abstract

As part of the adaptive immune system, T cells are critical to maintain immune homeostasis. T cells provide protective immunity by killing infected cells and combatting cancerous cells. To do so, T cells produce and secrete effector molecules, such as granzymes, perforin, and cytokines such as tumor necrosis factor α and interferon γ. However, in immune suppressive environments, such as tumors, T cells gradually lose the capacity to perform their effector function. One way T cell effector function can be enhanced is through genetic engineering with tools such as clustered regularly interspaced short palindromic repeat (CRISPR)/ CRISPR-associated protein 9 (Cas9). This protocol explains in a step-by-step fashion how to perform a controlled electroporation-based CRISPR experiment to enhance human T cell effector function. Of note, these steps are suitable for CRISPR-mediated genome editing in T cells in general and can thus also be used to study proteins of interest that do not influence T cell effector function.

Key words CRISPR, T cells, Genome editing, Cytokines, Flow cytometry, Single cell

1 Introduction

T cells are critical to maintain immune homeostasis and provide protective immunity by killing infected and cancer cells. To do so, T cells produce and secrete effector molecules, such as granzymes, perforin, and cytokines, including tumor necrosis factor (TNF)-α and interferon (IFN)-γ [1–5]. However, in immune suppressive environments such as tumors, T cells gradually lose the capacity to perform their effector function. T cell responses can be dampened through different mechanisms, such as the exploitation of inhibitory receptors, i.e., programmed cell death 1 (PD-1) [6, 7]. This results in the inhibition of T cell effector function and can ultimately lead to tumor immune escape.

Several strategies have been employed to optimize T cell effector function. For cancer treatment, cellular therapies with tumor-infiltrating lymphocytes (TILs) [8], or T cells that have been genetically engineered to express a T cell receptor (TCR) [9, 10] or a chimeric antigen receptor (CAR) [11] have shown

Vijay Kumar (ed.), *Immune Homeostasis: Methods and Protocols*, Methods in Molecular Biology, vol. 2782,
https://doi.org/10.1007/978-1-0716-3754-8_16,

promise. TILs are patient-derived T cells that have been reprogrammed and expanded ex vivo prior to reinfusion into the patient [8]. These TILs have been selected against tumor antigens in vivo and are highly suitable for tumors expressing non-shared (neo-) antigens. In contrast, TCR-engineered T cells or CAR-T cells are engineered to be equipped with a (chimeric antigen) receptor that targets a defined tumor antigen, such as NY-ESO or CD19 [12–14]. Both strategies have shown remarkable therapeutic potential [8, 15].

The efficacy of T cell therapy can still be improved, for instance, through genetic engineering [16]. One gene editing tool that is suitable for the enhancement of T cell effector function is the clustered regularly interspaced short palindromic repeat (CRISPR) and CRISPR-associated (Cas) system [17–20]. This can be achieved by, e.g., ablation of inhibitory receptor expression [17, 18, 21] or by modulating pathways instrumental for cytokine production [19–22]. The diverse ways through which CRISPR can be used to enhance T cell effector function has recently been reviewed elsewhere [16].

Three components are required for successful CRISPR/Cas9-mediated genome editing: the Cas9 DNA nuclease with catalytic activity, a targeting CRISPR RNA (crRNA) that guides the Cas9 protein to the locus of interest, and a trans-activating CRISPR RNA (tracrRNA) that facilitates Cas9 protein activation upon reaching the locus of interest [23]. Another requirement of Cas9-mediated genome cleavage is the presence of a protospacer-adjacent motif, or PAM, directly adjacent to the crRNA site [24]. The PAM is a 3–8 base pair DNA sequence that is specific to the Cas protein employed [23, 25]. This variety in PAM sequences allows for a broad selection of cleavage sites that can be used to target virtually any DNA region of interest.

Several delivery methods have been employed to introduce the CRISPR/Cas9 system into (human) cells, but only (lenti)viral delivery and electroporation-based strategies have been used successfully to genetically engineer T cells [20, 22, 26–28]. CRISPR/Cas9-mediated genome editing can also be used together with TCR engineering, equipping T cells with a TCR directed against an antigen of interest, e.g., a tumor antigen [20]. In fact, CRISPR-mediated PD-1 knockout T cells with a transgenic NY-ESO-1 TCR generated through viral delivery of the Cas9 system have been used in several human trials [14, 29]. Recently, an all-in-one platform employing adeno-associated virus-mediated CRISPR/Cas9 and transgene delivery allows for simultaneous knockout and redirection of T cell antigen specificity [30, 31]. Additionally, a novel technique exploiting HIV-serotyping of Cas9 ribonuclear protein (RNP)-containing viral particles was shown to be suitable for the specific targeting of $CD4^+$ T cells in a mixed culture and was used for simultaneous genome editing and introduction of a TCR of interest [32]. This new tool in the CRISPR toolbox opens up new avenues of research and has therapeutic potential [33]. However, stable expression of gene-editing factors could be undesirable due to potential off-target effects [34] and could result in antibody-mediated responses directed against Cas9 proteins, as a majority of healthy donors, and cord blood units were shown to possess Cas9 antibodies [35, 36].

Therefore, electroporation-based CRISPR/Cas9 introduction might be most optimal for the enhancement of T cell effector function in view of the transient presence of Cas9 proteins. Cas9 can be introduced as a plasmid, mRNA, Cas9 protein, or Cas9 RNPs through electroporation [20, 23, 37–39]. Of note, T cell electroporation is potentially cytotoxic, i.e., resulting in up to 95% loss of, e.g., viable T cells in the first 24 h after electroporation [40]. However, T cells retain their expansion potential after electroporation-based CRISPR/Cas9-mediated genome editing [20], thus overcoming this initial set back.

This protocol explains in a step-by-step fashion how to perform a controlled electroporation-based CRISPR experiment to enhance human T cell effector function via Cas9 RNP-mediated genome editing. Of note, these steps are suitable for CRISPR-mediated genome editing in T cells in general and can thus also be used to study proteins of interest that do not influence T cell effector function.

2 Materials

2.1 T Cell Activation

1. (Cryopreserved) human peripheral blood mononuclear cells (PBMCs).
2. Iscove's modified Dulbecco's medium (IMDM), supplemented with 5% heat-inactivated fetal calf serum (FCS), 5% human serum, 2 mM L-glutamine, 20 U/mL penicillin G sodium salts, and 20 μg/mL streptomycin sulfate salts. After the removal of cells from αCD3 + αCD28 stimulation, the medium in which cells are cultured should additionally be supplemented with 100 IU/mL recombinant human interleukin (IL)-2 and 10 ng/mL recombinant human IL-15. Store at 4 °C.
3. Phosphate-buffered saline (PBS).
4. Anti-(α) murine immunoglobulin G (IgG), α cluster of differentiation (CD)3, and αCD28 (*see* **Note 1**).
5. 24-well plates.
6. Humidified incubator.
7. Centrifuge.

2.2 CRISPR-Mediated Genome Editing of Activated Human T Cells

1. Activated human T cells.
2. Medium containing IL-2 and IL-15 (prepared as described above), pre-warmed to 37 °C.
3. Phosphate-buffered saline (PBS).
4. Liquid nitrogen.
5. Targeting and non-targeting negative control crRNAs (*see* **Notes 2–5**).
6. tracrRNA (*see* **Note 3**).
7. Recombinant Cas9 protein, i.e., TrueCut Cas9 V2 (Thermo Fisher Scientific) (*see* **Note 6**).
8. DNA Lo-binding Eppendorf tubes.
9. P2 primary cell 4D-nucleofector X kit (Lonza), containing P2 primary cell nucleofector solution, supplement 1, pMaxGFP vector, 100 μL nucleocuvette vessels, and single-use pipettes (*see* **Note 7**).
10. Nuclease-free PCR tubes.
11. 48-well plates
12. 4D nucleofector X unit (Lonza)
13. Centrifuge.
14. Thermocycler.
15. Humidified incubator.

2.3 Validation of Genome Editing

1. 1×10^6 snap-frozen genome-edited cells
2. Lysis buffer, containing 50 mM Tris-HCl, 2.5 mM ethylenediaminetetraacetic acid (EDTA), 50 mM KCl, and 0.45% tergitol, at pH 8.0, freshly supplemented with 1 μg/mL proteinase K (Roche).
3. Locus-specific primers flanking crRNA site (*see* **Note 2**).
4. DreamTaq HotStart Green Polymerase (Thermo Fisher Scientific).
5. 0.8 V% agarose gel with SYBR safe (or equivalent)
6. PCR clean-up kit (Macherey Nagel).
7. DNA Lo-binding Eppendorf tubes.
8. Centrifuge.
9. Gel electrophoresis unit.
10. Rotating shaker with heat block.

2.4 Functional T Cell Assays via Flow Cytometry

1. Genome-edited T cells.
2. Medium.
3. Stimuli (i.e., αCD3 + αCD28, PMA + ionomycin, target cells expressing the antigen, or loaded with antigen, *see* **Note 8**).

4. Brefeldin A.
5. FACS buffer: PBS supplemented with 1% FCS and 2 mM EDTA. Store at 4 °C.

   Any left-over buffer can be used for subsequent experiments.
6. Cell surface antibodies, i.e., for identifying T cell populations (*see* **Note 9**).
7. Non-permeable dye for identifying dead cells (*see* **Note 9**).
8. Fixable live/dead dye (i.e., Near-IR Dead Cell Stain kit, Thermo Fisher Scientific).
9. Cytofix/Cytoperm Fixation/Permabilization kit (BD Biosciences), containing Fixation/Permeabilization solution and 10x Perm/Wash solution.
10. 1x Perm/Wash solution: Dilute BD Perm/Wash™ buffer 1:10 in water. Store at 4 °C.

    Any left-over buffer can be used for subsequent experiments.
11. Intracellular protein antibodies (*see* **Note 10**).
12. 96-well F-bottom and/or V-bottom plates.
13. Flow cytometer.
14. Plate centrifuge.
15. Thermocycler.

3 Methods

3.1 T Cell Activation

Perform all steps in a biosafety cabinet with sterile equipment and reagents.

Preparation

1. Coat sterile 24-well plate(s) with 0.5 μg/mL α murine IgG in PBS per well by adding 400 μL per well and incubating overnight at 4 °C (*see* **Note 1**).

   The number of wells that are required depend on experimental design and the amount of samples to be electroporated (see **Note 11***).*
2. Remove α murine IgG from wells by pipetting and wash wells once with PBS.
3. Coat wells with 1 μg/mL αCD3 in PBS by adding 400 μL per well and incubate for at least 2 h at 37 ° C + 5% CO_2 in a humidified incubator.
4. Remove αCD3 from wells by pipetting and wash wells once with 1 mL PBS per well.

Experimental Procedure

5. If cryopreserved human PBMCs are used, thaw cells according to standard operating procedure. For example, thaw cells by letting the frozen cells thaw until only a small clump remains. Transfer the nearly thawed cells to a sterile 50 mL tube by pouring. Next, add 25 mL medium in a dropwise fashion. Rest cells for ~1 h at 37 °C + 5% CO_2 in a humidified incubator. After resting, spin cells down by centrifugation at 400× *g* for 5 min at maximum acceleration and brake.

   Please note that all subsequent centrifugation steps are performed at maximum acceleration and break unless specified otherwise.

   Discard the supernatant and resuspend cell pellet in a proper volume for counting cells. Count cells and bring to a concentration of 1×10^6 cells/mL in the medium.

   If freshly isolated human PBMCs are used, count cells and bring to a concentration of 1×10^6 cells/mL in medium.

6. Add 1 mL of cell suspension (containing 1×10^6 PBMCs) to each well of the αCD3-coated 24-well plate and incubate cells for 72 h at 37 °C + 5% CO_2 in a humidified incubator.

3.2 CRISPR-Mediated Genome Editing of Activated Human T Cells

Perform all steps in a biosafety cabinet with sterile equipment and reagents.

Preparation

1. If required, reconstitute crRNA and tracrRNA at 100 μM (*see* **Notes 2–5**).
2. Prepare a sterile 48-well plate in which the cells will be cultured after electroporation: add 500 μL medium containing IL-2 and IL-15 per condition/well to the 48-well plate(s) and pre-warm to 37 °C in a humidified incubator.
3. Prepare additional required medium (at least 100 μL for each condition to be tested and controls) in a sterile vessel and pre-warm to 37 °C in a humidified incubator.
4. Reconstitute sufficient P2 buffer according to manufacturer's instructions (Lonza) (*see* **Note 7**).

Experimental Procedure

1. Mix oligos at equimolar ratios in nuclease-free PCR tubes. Per sample to be electroporated, the author typically combines 4.5 μL crRNA and 4.5 μL tracrRNA. If using two crRNAs in one condition, 2.25 μL of each crRNA is added to 4.5 μL tracrRNA.
2. Denature oligo mixture by heating at 95 °C for 5 min in a thermocycler.

3. Remove the oligo mixture from thermocycler and allow to cool down to room temperature. If employing fluorescently labeled oligos, protect the mixture from light as much as possible from this step onward until after electroporation.
4. Produce Cas9 RNPs by adding 30 μg Cas9 protein per sample to the oligo mixture and pipetting up and down gently three times (*see* **Note 6**).

   This results in the production of RNPs with a ratio of 30 μg Cas9 : 4.5 μL crRNA : 4.5 μL tracrRNA.
5. Incubate Cas9 RNPs for at least 10 min at room temperature.
6. Harvest activated T cells (72 h after αCD3/αCD28 activation) by gently but thoroughly resuspending the cells that have thoroughly attached to the bottom of the 24-well plates by pipetting 5–10 times. If not all cells have been removed after the first harvest, harvest a second time with cold PBS.
7. Spin down the cells through centrifugation at 400× *g* for 5 min.
8. Discard supernatant and resuspend the pellet in PBS. Count cells and bring to a concentration of 1×10^6 cells/mL in PBS.
9. Transfer 1 mL of cell suspension (containing 1×10^6 PBMCs) per condition to be tested to sterile DNA Lo-binding Eppendorf tubes.
10. Spin down the cells through centrifugation at 1200× *g* for 6 min. Completely remove supernatant.
11. Gently resuspend the cell pellet in 20 μL P2 buffer per Eppendorf tube by pipetting (*see* **Note 12**).
12. Add Cas9 RNPs to the samples, gently resuspend by pipetting and incubate for 2 min.
13. Transfer cells to 16-well strip nucleocuvette by pipetting gently, and as directly on the bottom of the cuvette as possible. Limit bubble formation while pipetting, as bubbles can interfere with electroporation procedure. If required, gently tap nucleocuvette vessels trip on lab bench for bubbles located at the bottom of the strip to float to the surface.
14. Electroporate cells in a 4D nucleofector X unit with program EH-100.
15. Add 100 μL of pre-warmed medium dropwise to each cuvette containing cell suspension and allow cells to rest for 5 min at 37 ° C + 5% CO_2 in a humidified incubator.
16. After resting, transfer cells to pre-prepared 48-well plate wells with pre-warmed medium via the disposable pipette supplied with the nucleofection kit by adding cell mixture dropwise to the 48-well plate well.
17. Culture cells at 37 °C + 5% CO_2 in a humidified incubator and split cells as required (*see* **Note 5**).

3.3 Validation of Genome Editing Through Agarose Gel Analysis or Sanger Sequencing

Steps 1 and 2 should be performed in a biosafety cabinet with sterile equipment and reagents. After snap-freezing the cells in step 3, the rest of the protocol can be performed on a lab bench outside a biosafety cabinet with clean equipment and reagents.

1. Harvest 1×10^6 cells per condition to be tested. This can be performed at the earliest 48 h after electroporation.
2. Wash cells twice with PBS by centrifuging cells at 400× *g* for 5 min. Completely remove supernatant between washes.
3. Snap-freeze the cell pellets in liquid nitrogen. Pellets can be immediately used or stored at −20 °C for genomic DNA isolation at a later time point.
4. Incubate cell pellets overnight in 50 μL lysis buffer at 56 °C while rotating at 850 RPM.
5. After overnight incubation, deactivate proteinase K by incubating for 15 min at 95 °C while rotating at 850 RPM.
6. Centrifuge cell lysate at 20.000× *g* for 10 min.
7. Transfer the supernatant containing genomic DNA to a new DNAse free tube. PCR for genomic loci can be performed now, or DNA can be stored at −20 °C for analysis at a later time point.
8. Amplify locus of interest via PCR using DreamTaq HotStart Green Polymerase according to manufacturer's instructions (Thermo Fisher Scientific), or employ a polymerase of choice according to SOP.
9. If employing 2 crRNAs that result in a > ~50 base pair deletion, analyze PCR fragments on 0.8 V% agarose gel. If employing 1 crRNA, perform Sanger sequencing according to standard operating procedure after purifying the PCR product with a PCR clean-up kit according to manufacturer's instruction (Machery Nagel). Genome editing can be assessed by visual inspection of the agarose gel or Sanger results [20] or by making use of genome analysis algorithms such as Tracking of Indels by DEcomposition (TIDE) [41].

3.4 Functional T Cell Assays via Flow Cytometry

Perform all steps in a biosafety cabinet with sterile equipment and reagents.

1. Prepare stimuli (*see* **Note 8**). In case of plate-bound αCD3/αCD28 stimulation, the preparation procedure described in Subheading 3.1, starting at **step 2**, can be followed to coat a sterile 96-well flat-bottom plate with 50 μL of αCD3 in PBS per well.
2. Harvest cells as described above. Spin down the cells through centrifugation at 400× *g* for 5 min.

3. Discard supernatant and resuspend the pellet in 1 mL medium. Count cells and bring to a concentration of 1×10^6 cells/mL in PBS.
4. Activate 1×10^5 cells in 200 μL medium containing the stimulus of choice and 1 μg/mL Brefeldin A in a suitable sterile 96-well plate, i.e., by combining 100 μL cell suspension with 2× Brefeldin A with 100 μL medium with 2× concentrated stimuli. Cover the plate with a lid and culture cells for the desired amount of time in a humidified incubator at 37 °C + 5% CO_2 (*see* **Note 13**).
5. Remove plate(s) from incubator. Transfer cells to a 96-well V-bottom plate and centrifuge at 650× *g* for 3 min at 4 °C. Remove supernatant.
6. Add desired surface antibodies in 30 μL FACS buffer and resuspend by pipetting. Incubate for 20 min at 4 °C in the dark (*see* **Note 9**).
7. After incubation, add 150 μL FACS buffer and centrifuge at 650× *g* for 3 min at 4 °C. Remove supernatant.
8. Add 50 μL Fixation/permeabilization solution. Resuspend the cell pellet either by briefly vortexing the plate, or pipetting, and incubate for 20–30 min at 4 °C in the dark.
9. After incubation, add 150 μL pre-treated 1x Perm/Wash solution to the cells and centrifuge at 650× *g* for 3 min at 4 °C. Remove supernatant.
10. Add desired intracellular antibodies in 30 μL 1x Perm/Wash solution, and resuspend cell pellets by pipetting. Incubate for 30 min at 4 °C in the dark (*see* **Note 10**).
11. After incubation, wash cells twice by adding 150 μL 1x Perm/Wash solution to the cells and centrifuging at 650× *g* for 3 min at 4 °C. Remove supernatant.
12. Resuspend cells in 100 μL Perm/Wash solution and measure samples on a suitable flow cytometer.

4 Notes

1. For primary T cell activation, pre-coating plates with α murine IgG prior to coating with αCD3 in general results in optimal T cell activation due to enhanced antibody presentation.
2. For designing crRNAs and genomic PCR primers, the author has used the Benchling CRISPR tool (available online at https://www.benchling.com/). Preferably, high-scoring crRNA sequences are selected, and potential off-target effects can be assessed via interrogation of the human genome through, e.g., BlastN (NCBI).

3. The author has previously successfully used custom-designed crRNAs and tracrRNAs from Integrated DNA Technologies for CRISPR-mediated genomic editing to enhance T cell effector function [20, 22]. Other primers, e.g., targeting PD-1, have also been described in the literature. An overview of some of the primers that could be employed to enhance T cell effector function can be found in Table 1.
4. As a positive control, the author has used *IFNG*-targeting crRNAs that have been previously described [28]. As a negative control, the author has used commercially available negative control crRNA 1 (Integrated DNA Technologies) [20]. Other negative control crRNAs have also been described [42, 48].
5. To facilitate confirmation of successful electroporation, the author has previously used fluorescently labeled probes, i.e., ATTO550-labeled tracrRNA (Integrated DNA Technologies) that can be measured via flow cytometry [20].
6. The author has successfully used TrueCut Cas9 V2 (Thermo Fisher Scientific) [20] and GeneArt Platinum Cas9 (Invitrogen, data not shown) for genomic modification of human T cells.
7. According to the manufacturer's protocol, reconstituted P2 buffer can be stored for 1 month at 4 °C. Therefore, reconstitute only sufficient buffer for each experiment to prolong shelf life.
8. For T cell restimulation, 10 ng/mL PMA + 1 μM ionomycin (both Sigma Aldrich) [49], 1 μg/mL αCD3 and 1 μg/mL αCD28 [20, 49], or between 1 and 5 μg/mL peptide (pools) [49], or a stimuli of interest can be used.
9. Typically, author uses antibodies directed against CD4 (clone SK3, i.e., BUV737-labeled [BD Biosciences]) and CD8 (clone SK1, i.e., BUV805-labeled [BD Biosciences]) surface antigens for the identification of T cell subsets, and a near/IR dye (Thermo Fisher Scientific) to identify dead cells [20].
10. Typically, author uses antibodies directed against IFN-γ (4S.B3 i.e., PE-labeled [BioLegend]), IL-2 (clone MQ1-17H12, i.e., AF488-labeled [BioLegend], and TNF-α (clone MAb11, i.e., PE-Cy7-labeled [BD Biosciences]) [20].
11. Besides experimental conditions, author employs the following controls: non-targeting control (i.e., non-targeting crRNA + Cas9 RNPs), Cas9 only, tracrRNA only, and a non-nucleofected sample that has been incubated in P2 similar to all other conditions.
12. P2 buffer is toxic to cells. All subsequent steps should be executed carefully but swiftly. If targeting a large amount of proteins, consider performing electroporation in two batches

Table 1
Previously reported CRISPR crRNA primers used to enhance human T cell effector function

Target	crRNA Primer (5′ -> 3′)	References
CTLA4	CCAGGTGACTGAAGTCTGTG	[28]
	CACTGTCACCCGGACCTCAG	[28]
	GCACGTGGCCCAGCCTGCTG	[28]
	CTCAGCTGAACCTGGCTACC	[42, 43]
	CCTTGGATTTCAGCGGCACA	[42, 43]
	GCACAAGGCTCAGCTGAACC	[42, 43]
DGKA	CTCTCAAGCTGAGTGGGTCC	[44]
DGKZ	ACGAGCACTCACCAGCATCC	[44]
IFNG 3′UTR	GTGAGACAGTCACAGGATAT	[20]
	ATTTAAAATTCAAATATTGC	
LAG3	ATGTGGGAGGCTCAGTTCCT	[45]
	GCTGCAGAAACAGCAAGCCC	[45]
	TGCTGTTTCTGCAGCCGCTT	[45]
	GCTGTTTCTGCAGCCGCTTT	[45]
	GTTTCTGCAGCCGCTTTGGG	[45]
PD1	CTACAACTGGGCTGGCGGCC	[17, 38, 46]
	CGACTGGCCAGGGCGCCTGT	[17, 38, 46]
	CTGCAGCTTCTCCAACACAT	[17, 38, 46]
	CGTGTCACACAACTGCCCAA	[17, 38, 46]
	GCCCTGGCCAGTCGTCTGGG	[17, 38, 46]
	GTCTGGGCGGTGCTACAACT	[44]
	GGCGCCCTGGCCAGTCGTCTGGG	[47]
	AGGCGCCCTGGCCAGTCGTC	[48]
	CGTCTGGGCGGTGCTACAAC	[48]
PTPN2	AAGGAGUUACAUCUUAACAC	[46]
	CAGUUUAGUUGACAUAGAAG	[46]
TGFBR2	CCTGAGCAGCCCCCGACCCA	[46]
TIGIT	CAGGCACAATAGAAACAACG	[28]
	GACCTGGGTCACTTGTGCCG	[28]
	CAGCGCATGCTTCTGCCCAG	[28]
ZC3HAV1	AAAATCCTGTGCGCCCACGG	[22]
	GTCTCTGGCAGTACTTGCGA	[22]
ZFP36L1	AAACGGTGCCTGTAAGTACG	[22]
	GTCTCGCGAGCTCAGAGCGG	[22]

and keeping the cells for batch 2 in PBS as described in Subheading 3.2, **step 8** until finished with the first batch.

13. Typically, author re-stimulates T cells for up to 8 h with αCD3/αCD28, with 4- or 6-h stimulations being most common [20]. If required, snapshots of cytokine production can be taken by adding Brefeldin A only during the last 2 h of stimulation [20].

References

1. Pennock ND, White JT, Cross EW et al (2013) T cell responses: naïve to memory and everything in between. Adv Psyiol Educ 37:273–283. https://doi.org/10.1152/advan.00066.2013
2. Voskoboinik I, Whisstock JC, Trapani JA (2015) Perforin and granzymes: function, dysfunction and human pathology. Nat Rev Immunol 15:388–400. https://doi.org/10.1038/nri3839
3. Golubovskaya V, Wu L (2016) Different subsets of T cells, memory, effector functions, and CAR-T immunotherapy. Cancers (Basel) 8. https://doi.org/10.3390/cancers8030036
4. Salerno F, Freen-van Heeren JJ, Guislain A et al (2019) Costimulation through TLR2 drives Polyfunctional CD8$^+$ T cell responses. J Immunol 202:714–723. https://doi.org/10.4049/jimmunol.1801026
5. Nicolet BP, Guislain A, van Alphen FPJ et al (2020) CD29 marks superior cytotoxic human CD8+ T cells. Proc Natl Acad Sci 117:6686–6696. https://doi.org/10.1101/562512
6. Seidel JA, Otsuka A, Kabashima K (2018) Anti-PD-1 and anti-CTLA-4 therapies in cancer: mechanisms of action, efficacy, and limitations. Front Oncol 8. https://doi.org/10.3389/fonc.2018.00086
7. Tang F, Zheng P (2018) Tumor cells versus host immune cells: whose PD-L1 contributes to PD-1/PD-L1 blockade mediated cancer immunotherapy? Cell Biosci 8:34. https://doi.org/10.1186/s13578-018-0232-4
8. Rohaan MW, Van Den BJH, Kvistborg P, Haanen JBAG (2018) Adoptive transfer of tumor-infiltrating lymphocytes in melanoma: a viable treatment option. J Immunother Cancer 6:102
9. Albers JJ, Ammon T, Gosmann D et al (2019) Gene editing enables T-cell engineering to redirect antigen specificity for potent tumor rejection. Life Sci Alliance 2:10.26508/lsa.201900367
10. Morton LT, Reijmers RM, Wouters AK et al (2020) Simultaneous deletion of endogenous TCRαβ for TCR gene therapy creates an improved and safe cellular therapeutic. Mol Ther 28:64–74. https://doi.org/10.1016/j.ymthe.2019.10.001
11. Eyquem J, Mansilla-soto J, Giavridis T et al (2017) Targeting a CAR to the TRAC locus with CRISPR/Cas9 enhances tumour rejection. Nature 543:113–117. https://doi.org/10.1038/nature21405.Targeting
12. Pettitt D, Arshad Z, Smith J et al (2018) CAR-T cells : a systematic review and mixed methods analysis of the clinical trial landscape. Mol Ther 26:342–353. https://doi.org/10.1016/j.ymthe.2017.10.019
13. Weinkove R, George P, Dasyam N, McLellan AD (2019) Selecting costimulatory domains for chimeric antigen receptors: functional and clinical considerations. Clin Transl Immunol 8:e1049. https://doi.org/10.1002/cti2.1049
14. Stadtmauer EA, Fraietta JA, Davis MM et al (2020) CRISPR-engineered T cells in patients with refractory cancer. Science (80-) 367:10.1126/science.aba7365
15. Mohanty R, Chowdhury CR, Arega S et al (2019) CAR T cell therapy: a new era for cancer treatment. Oncol Rep 42:2183–2195. https://doi.org/10.3892/or.2019.7335
16. Freen-van Heeren JJ (2021) Using CRISPR to enhance T cell effector function for therapeutic applications. Cytokine X 3:100049. https://doi.org/10.1016/j.cytox.2020.100049
17. Rupp LJ, Schumann K, Roybal KT et al (2017) CRISPR/Cas9-mediated PD-1 disruption enhances anti-tumor efficacy of human chimeric antigen receptor T cells. Sci Rep 7:1–10. https://doi.org/10.1038/s41598-017-00462-8
18. Hu W, Zi Z, Jin Y et al (2019) CRISPR/Cas9-mediated PD-1 disruption enhances human mesothelin-targeted CAR T cell effector functions. Cancer Immunol Immunother 68:365–377. https://doi.org/10.1007/s00262-018-2281-2
19. Wei J, Long L, Zheng W et al (2019) Targeting REGNASE-1 programs long-lived effector T cells for cancer therapy. Nature 576:471–476. https://doi.org/10.1038/s41586-019-1821-z
20. Freen-van Heeren JJ, Popović B, Guislain A, Wolkers MC (2020) Human T cells employ conserved AU-rich elements to fine-tune IFN-γ production. Eur J Immunol 50:949–958. https://doi.org/10.1002/eji.201948458
21. Freen-van Heeren JJ (2021) Post-transcriptional control of T-cell cytokine production: implications for cancer therapy. Immunology 164:57. https://doi.org/10.1111/imm.13339
22. Popovic B, Guislain A, Engels S et al (2023) Time-dependent regulation of cytokine production by RNA binding proteins defines T cell effector function. Cell Rep 42:112419. https://doi.org/10.1016/j.celrep.2023.112419

23. Ran FA, Hsu PD, Wright J et al (2013) Genome engineering using the CRISPR-Cas9 system. Nat Protoc 8:2281–2308. https://doi.org/10.1038/nprot.2013.143
24. Jinek M, Chylinski K, Fonfara I et al (2012) A programmable dual-RNA-guided DNA endonuclease in adaptive bacterial immunity. Science (80-) 337:816–821. https://doi.org/10.1126/science.1225829
25. Adli M (2018) The CRISPR tool kit for genome editing and beyond. Nat Commun 9: 1911. https://doi.org/10.1038/s41467-018-04252-2
26. Wang W, Ye C, Liu J et al (2014) CCR5 gene disruption via Lentiviral vectors expressing Cas9 and single guided RNA renders cells resistant to HIV-1 infection. PLoS One 1–26. https://doi.org/10.1371/journal.pone.0115987
27. Li C, Guan X, Du T et al (2015) Inhibition of HIV-1 infection of primary CD4+ T-cells by gene editing of CCR5 using CRISPR/Cas9. J Gen Virol 96:2381–2393. https://doi.org/10.1099/vir.0.000139
28. Seki A, Rutz S (2018) Optimized RNP transfection for highly efficient CRISPR/Cas9-mediated gene knockout in primary T cells. J Exp Med 215:985–997. https://doi.org/10.1084/jem.20171626
29. Lu Y, Xue J, Deng T et al (2020) Safety and feasibility of CRISPR-edited T cells in patients with refractory non-small-cell lung cancer. Nat Med 26:732–740. https://doi.org/10.1038/s41591-020-0840-5
30. Dai X, Park JJ, Du Y et al (2023) Massively parallel knock-in engineering of human T cells. Nat Biotechnol 41:1239. https://doi.org/10.1038/s41587-022-01639-x
31. Opstelten R, Freen-Van Heeren J (2023) CLASH of the titans: how CAR-T cells can triumph over tumors. Cris J 6:87–89. https://doi.org/10.1089/crispr.2023.0003
32. Hamilton JR, Tsuchida CA, Nguyen DN et al (2021) Targeted delivery of CRISPR-Cas9 and transgenes enables complex immune cell engineering. Cell Rep 35:109207. https://doi.org/10.1016/j.celrep.2021.109207
33. Freen-van Heeren JJ (2021) Exploiting HIV-1 tropism to target CD4+ T cells for CRISPR. Immunol Cell Biol 99:677–679. https://doi.org/10.1111/imcb.12487
34. Kim S, Kim D, Cho S et al (2014) Highly efficient RNA-guide genome editing. Genome Res 24:1012–1019. https://doi.org/10.1101/gr.171322.113.Freely
35. Simhadri VL, McGill J, McMahon S et al (2018) Prevalence of pre-existing antibodies to CRISPR-associated nuclease Cas9 in the USA population. Mol Ther – Methods Clin Dev 10:105–112. https://doi.org/10.1016/j.omtm.2018.06.006
36. Charlesworth CT, Deshpande PS, Dever DP et al (2019) Identification of preexisting adaptive immunity to Cas9 proteins in humans. Nat Med 25:249–254. https://doi.org/10.1038/s41591-018-0326-x
37. Liang X, Potter J, Kumar S et al (2015) Rapid and highly efficient mammalian cell engineering via Cas9 protein transfection. J Biotechnol 208:44–53. https://doi.org/10.1016/j.jbiotec.2015.04.024
38. Schumann K, Lin S, Boyer E et al (2015) Generation of knock-in primary human T cells using Cas9 ribonucleoproteins. Proc Natl Acad Sci 112:10437–10442. https://doi.org/10.1073/pnas.1512503112
39. Kornete M, Marone R, Jeker LT (2018) Highly efficient and versatile plasmid-based gene editing in primary T cells. J Immunol 200:2489–2501. https://doi.org/10.4049/jimmunol.1701121
40. Chicaybam L, Sodre AL, Curzio BA, Bonamino MH (2013) An efficient low cost method for gene transfer to T lymphocytes. PLoS One 8:1–11. https://doi.org/10.1371/journal.pone.0060298
41. Brinkman EK, Chen T, Amendola M, Van Steensel B (2014) Easy quantitative assessment of genome editing by sequence trace decomposition. Nucleic Acids Res 42:1–8. https://doi.org/10.1093/nar/gku936
42. Zhang W, Shi L, Zhao Z et al (2019) Disruption of CTLA-4 expression on peripheral blood CD8+ T cell enhances anti-tumor efficacy in bladder cancer. Cancer Chemother Pharmacol 83:911–920. https://doi.org/10.1007/s00280-019-03800-x
43. Shi L, Meng T, Zhao Z et al (2017) CRISPR knock out CTLA-4 enhances the anti-tumor activity of cytotoxic T lymphocytes. Gene 636:36–41. https://doi.org/10.1016/j.gene.2017.09.010
44. Jung IY, Kim YY, Yu HS et al (2018) CRISPR/Cas9-mediated knockout of DGK improves antitumor activities of human T cells. Cancer Res 78:4692–4703. https://doi.org/10.1158/0008-5472.CAN-18-0030
45. Zhang Y, Zhang X, Cheng C et al (2017) CRISPR-Cas9 mediated LAG-3 disruption in CAR-T cells. Front Med 11:554–562. https://doi.org/10.1007/s11684-017-0543-6
46. Tang N, Cheng C, Zhang X et al (2020) TGF-β inhibition via CRISPR promotes the long-term efficacy of CAR T cells against solid

tumors. JCI Insight 5:e133977. https://doi.org/10.1172/jci.insight.133977

47. Zhang C, Peng Y, Hublitz P et al (2018) Genetic abrogation of immune checkpoints in antigen-specific cytotoxic T-lymphocyte as a potential alternative to blockade immunotherapy. Sci Rep 8:5549. https://doi.org/10.1038/s41598-018-23803-7

48. Zhao Z, Shi L, Zhang W et al (2018) CRISPR knock out of programmed cell death protein 1 enhances anti-tumor activity of cytotoxic T lymphocytes. Oncotarget 9:5208–5215. https://doi.org/10.18632/oncotarget.23730

49. Nicolet BP, Guislain A, Wolkers MC (2017) Combined single-cell measurement of cytokine mRNA and protein identifies T cells with persistent effector function. J Immunol 198:962–970. https://doi.org/10.4049/jimmunol.1601531

Chapter 17

Assessing Antigen-Specific T Cell Responses Through IFN-γ Enzyme-Linked Immune Absorbent Spot (ELISpot)

Julian J. Freen-van Heeren, Virginia Palomares Cabeza, David Cobeta Lopez, Domenique Kivits, Irma Rensink, Annelies W. Turksma, and Anja ten Brinke

Abstract

T cells are instrumental in protecting the host against invading pathogens and the development of cancer. To do so, they produce effector molecules such as granzymes, interleukins, interferons, and perforin. For the development and immunomonitoring of therapeutic applications such as cell-based therapies and vaccines, assessing T cell effector function is paramount. This can be achieved through various methods, such as ^{51}Cr release assays, flow cytometry, and enzyme-linked immune absorbent spot (ELISpot) assays. For T cell ELISpots, plates are coated with antibodies directed against the effector molecule of interest (e.g., IFN-g). Subsequently, peripheral blood mononuclear cells (PBMCs) or isolated T cells are cultured on the plate together with stimuli of choice, and the production of effector molecules is visualized via labeled detection antibodies. For clinical studies, ELISpot is currently the gold standard to determine antigen-specific T cell frequencies. In contrast to ^{51}Cr release assays, ELISpot allows for the exact enumeration of responding T cells, and compared to flow cytometry, ELISpot is more cost-effective and high throughput. Here, we optimize and describe, in a step-by-step fashion, how to perform a controlled IFN-γ ELISpot experiment to determine the frequency of responding or antigen-specific T cells in healthy human volunteers. Of note, this protocol can also be employed to assess the frequency of antigen-specific T cells induced in, e.g., vaccination studies or present in cellular products.

Key words ELISpot, T cells, Effector function, IFN-g

1 Introduction

T cells are important for protection against pathogens and to kill malignantly transformed cancer cells. To do so, they produce effector molecules, including cytolytic molecules such as granzymes and perforin, and cytokines such as interferon (IFN)-γ, interleukins (ILs), and tumor necrosis factor (TNF)-α [1–9]. Therefore,

Authors Annelies W. Turksma and Anja ten Brinke shared senior authorship.

Vijay Kumar (ed.), *Immune Homeostasis: Methods and Protocols*, Methods in Molecular Biology, vol. 2782,
https://doi.org/10.1007/978-1-0716-3754-8_17,

monitoring antigen-specific T cell activation is of prime importance in, e.g., therapeutic settings such as cancer therapy and vaccination studies.

Several techniques are readily available for assessing the production of T cell effector molecules after (antigen-specific) stimulation. Several of these techniques, such as enzyme-linked immunosorbent assay (ELISA) or bead-based measurements, assess the production of effector molecules such as IFN-γ in the supernatant [10]. While this allows for the quantification of the amount of produced effector molecules, it does not provide information at a single-cell level. Other techniques are more suitable for this. For instance, flow cytometry allows for in-depth cellular phenotyping and/or simultaneous assessment of multiple effector molecules [8, 11–14]. However, flow cytometry also requires reasonably high cell numbers, especially when assessing cells present at a low frequency, such as antigen-specific T cells. Furthermore, flow cytometric assays and analyses are time-consuming and can be very costly.

Another technique that provides single-cell resolution is the enzyme-linked immune absorbent spot (ELISpot) assay. For T cell ELISpot, the membrane surface of a 96-well plate is coated with a detection antibody directed against the effector molecule(s) of interest (e.g., IFN-γ) (see Fig. 1a). Isolated or cryopreserved peripheral blood mononuclear cells (PBMCs) are added, together with an antigen of choice. Alternatively, a mitogen (e.g., phytohemagglutinin [PHA]) or stimulatory antibody (e.g., anti-CD3) can be used to activate T cells. During overnight stimulation, activated T cells will secrete effector molecules, such as IFN-γ. The antibodies coated on the bottom of the well will capture the produced molecule. With a (labeled) detection antibody, the captured secreted protein can be visualized, resulting in the formation of a spot on the location where the, e.g., IFN-γ-producing T cell was present (see Fig. 1b). Of note, by making use of fluorescently labeled antibodies, several effector molecules can be visualized simultaneously [15]. This multiplex ELISpot is also known as FluoroSpot.

The advantage of using ELISpot compared to other techniques assessing T cell effector function is that, like flow cytometry [14], ELISpot allows for the enumeration of antigen-specific cells. In contrast, classical T cell effector assays such as the ^{51}Cr release assay, in which target cells are labeled with radioactive chromium and subsequently co-cultured with cytotoxic T cells to induce target cell killing, provide a bulk measurement. While target cell killing is seen as a true effector function, the secretion of cytotoxic molecules such as IFN-γ and granzyme B was shown to be more sensitive in picking up reactive T cells compared to a ^{51}Cr release assay, while also providing an exact enumeration of reactive T cells [16]. Furthermore, the ELISpot platform is readily standardized across multiple laboratories [17], making it extremely suitable for

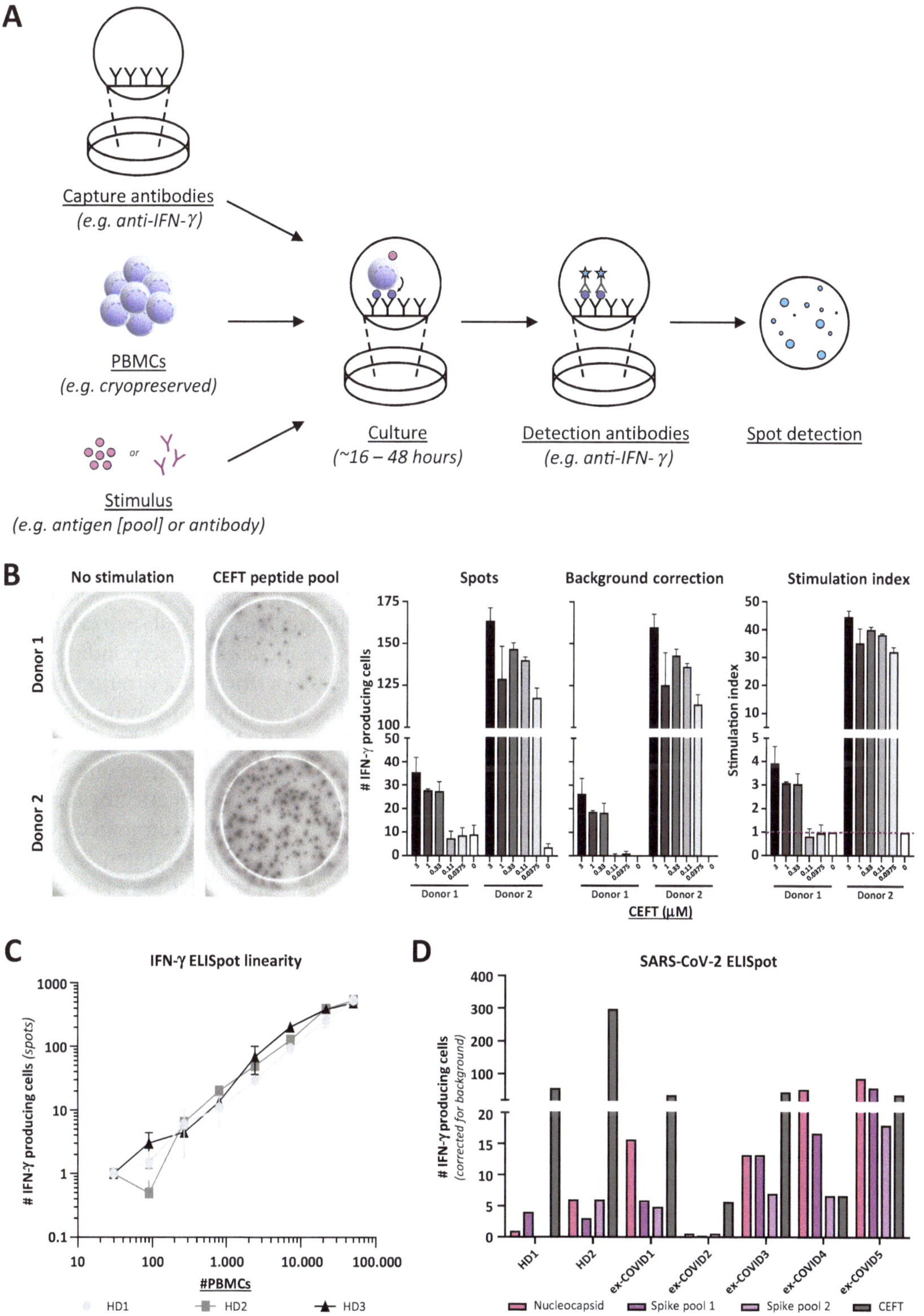

Fig. 1 IFN-γ ELISpot to measure (antigen-specific) T cell responses. (**a**) Schematic representation of an ELISpot assay. (**b**) Representative images (left) or summary graphs (right), depicting IFN-γ production. Cryopreserved PBMCs were rested overnight and stimulated overnight in AIMV medium with indicated amounts of CEFT peptide pool. Left graph indicates the number of spots counted, middle graph displays

use in clinical studies [18, 19]. Lastly, in contrast to flow-based assays, ELISpot is more cost-effective to perform in a high-throughput fashion.

Here, we present several examples of an ELISpot assay. For example, we show the production of IFN-γ by T cells from two healthy donors upon stimulation with a peptide pool containing peptides derived from the major antigens of CMV, EBV, influenza, and tetanus toxoid (CEFT). By performing antigen titrations, the optimal concentration of peptides or peptide pools can be determined. Typically, ELISpot data is presented as either total spots per 2×10^5 or 1×10^6 cells (see Fig. 1b, left graph), as total spots corrected for background (see Fig. 1b, middle graph) or as a ratio compared to the background spots, also known as the stimulation index (see Fig. 1b, right graph). When assessing donor responses, all three measurements can be employed. For comparing donors, using the total spots corrected for background or stimulation index corrects for any aspecific activation present in donors. In contrast, in the case of high background or low frequencies of antigen-specific T cells, using the stimulation index might skew the data. In those situations, using the total spots corrected for background is more informative.

ELISpot assays require few antigen-specific cells per sample, as it can be used to detect low frequencies of responding cells [20, 21]. As a proof of principle, we performed a titration experiment and showed that robust IFN-γ production can be observed from as little as 800 PHA-stimulated PBMCs (see Fig. 1c). Furthermore, ELISpot protocols are readily implemented and easy to adapt. Lastly, ELISpot-based assessment of T cell effector function is suitable for high-throughput analysis through automated evaluation [22] and is the gold standard for assessing antigen-specific T cells in, e.g., clinical trials and vaccination studies [23, 24]. Of note, also in healthy volunteers, ELISpot-based assessment of antigen-specific T cell frequencies can be informative, for instance, to assess whether an individual possesses cellular immunity toward a pathogen, as is exemplified in Fig. 1d, where we enumerated the amount

Fig. 1 (continued) the number of spots corrected for the average amount of spots in the background wells [0 mM CEFT, white bar], and the right graph displays the stimulation index, calculated as the number of spots divided by the average number of spots in the background wells [0 mM CEFT, white bar]. Data displayed in the summary graphs are mean ± standard deviation (SD) of three technical replicates of two healthy donors. (**c**) Summary graph of cryopreserved human PBMCs stimulated overnight with 10 mg/mL PHA in AIMV medium and measured via IFN-γ ELISpot. Data displayed in the summary graphs are the mean ± SD of three technical replicates of three healthy donors, representing at least two independently performed experiments. (**d**) Summary graph of cryopreserved human PBMCs stimulated overnight with 1 mM of indicated stimuli in AIMV medium and measured via IFN-γ ELISpot. Data displayed in the summary graphs are the mean number of spots corrected for the background production of three replicates of two healthy donors (HD) and five donors that have recovered from an infection with SARS-CoV-2 (indicated as "ex-COVID"), measured in two independently performed experiments

of severe acute respiratory syndrome coronavirus 2 (SARS-CoV-2)-specific cells in two healthy volunteers previously unexposed to SARS-CoV-2 to the best of their knowledge and five blood donors recovered from infection with SARS-CoV-2.

Here, we describe, in a step-by-step fashion, how to perform a controlled IFN-γ ELISpot experiment to determine the frequency of responding or antigen-specific T cells in healthy human volunteers. Typically, in immune monitoring of clinical trials, the production of IFN-γ is assessed. However, T cells produce a myriad of other cytokines and effector molecules as well, including, but not limited to, IL-2, IL-4, IL-5, IL-17, IL-21, TNF-α, granzymes, and perforins [8, 9, 25, 26]. These can also be assessed via ELISpot by using the general guidelines set out in this protocol, with optimization for the stimulus duration as each effector molecule is produced with individual kinetics. However, also other immune cells can produce these effector molecules, especially granzymes (e.g., produced by NK cells) and TNF-α (e.g., produced by monocytes). As previously mentioned, by making use of fluorescent dyes, multiple effector molecules can be assessed simultaneously via FluoroSpot [27, 28]. Especially when attempting to assess effector molecules produced by more cell subsets, analyzing double or triple producers can be beneficial over the assessment of just single, e.g., TNF-α production. Of note, this protocol can also be employed to assess the frequency of antigen-specific T cells induced in, e.g., vaccination studies or present in cellular products.

2 Materials

2.1 PBMC Cryopreservation

1. Human PBMCs (*see* **Note 1**).
2. Phosphate-buffered saline (PBS) (sterile).
3. Lymphoprep (e.g., Alere Technologies).
4. Medium: Iscove's modified Dublecco's medium (IMDM), 4% heat-inactivated fetal calf serum (FCS), 20 U/mL penicillin G sodium, and 20 μg/mL streptomycin sulfate – can be prepared in advance, stored at 4 °C until the day of use.
5. Freezing medium: IMDM, 8% heat-inactivated FCS, 10% sterile dimethylsulfoxide) – can be prepared in advance, stored at 4 °C until the day of use.
6. Wash buffer: PBS, 10% human serum albumin 40 g/L, 0.4% tri-sodium carbonate [TNC]) – can be prepared in advance, stored at 4 °C until the day of use.
7. Sterile 15 mL conical tubes.
8. Sterile 50 mL conical tubes.
9. (Micro) pipettes.
10. PipetBoy.
11. 80% ethanol spray.

2.2 PBMC Thawing and Overnight Resting

1. Medium (*see* **Note 2**).
2. (Cryopreserved) PBMCs
3. Sterile 50 mL conical tubes.
4. Cell counter.
5. Centrifuge.
6. Humidified incubator.
7. (Micro) pipettes.
8. Water bath.
9. 80% ethanol spray.

2.3 Coating ELISpot Plate

1. 35% ethanol in double-distilled H_2O: for one plate, 1.75 mL 100% ethanol or 2.19 mL 80% ethanol to 3.25 mL double-distilled H_2O (when using 100% ethanol) or 2.81 mL double-distilled H_2O (when using 80% ethanol) – needs to be prepared fresh on the day of coating.
2. IFN-γ ELISpot kit, MabTech, category #3420-2A.
3. PBS (sterile).
4. ELISpot plates.
5. (Micro) pipettes.
6. Multichannel pipette.
7. Tissue paper.
8. 80% ethanol spray.

2.4 T Cell Activation

1. Medium (*see* **Note 2**).
2. Rested human PBMCs (*see* **Note 3**).
3. T cell stimuli (*see* **Notes 4** and **5**).
4. Sterile 50 mL conical tubes.
5. Cell counter.
6. Centrifuge.
7. Humidified incubator.
8. (Micro) pipettes.
9. Multichannel pipette.

2.5 IFN-γ ELISpot

1. PBS
2. PBS, 0.5% FCS: Add 100 μL FCS to 19.9 mL PBS for a 0.5% solution
3. 0.2 μm filter
4. 0.45 μm filter
5. Electric fan (optional)
6. (Micro) pipettes
7. Multichannel pipette

3 Methods

3.1 PBMC Isolation (Cryopreservation – Optional)

Perform all steps in a biosafety cabinet with sterile equipment and reagents.

1. When starting with whole blood, combine all tubes from one donor in a T175 flask. When starting with buffy coats, transfer buffy coat to a T175 flask.
2. Dilute blood/buffy coat at a 1 to 1 ratio with sterile PBS. For example, add 50 mL PBS to 50 mL of whole blood/buffy coat.
3. Prepare sterile 50 mL tubes by adding 15 mL Lymphoprep per tube. Per tube, 20 mL of diluted blood/buffy coat can be added. For example, in case of 100 mL diluted blood/buffy coat; prepare five 50 mL tubes with 15 mL Lymphoprep/tube.
4. Gently layer 20 mL of diluted blood/buffy coat by pipetting at the slowest speed with a PipetBoy.
5. Perform Ficoll density separation by centrifugation at 1000× *g* for 15 min at room temperature, acceleration 4, no brake.
6. After centrifugation, remove the top 2 mL of serum containing most of the thrombocytes, and discard.
7. Collect the white interface from each 50 mL tube and transfer each interphase to a new 50 mL tube.
8. Add 45 mL of wash buffer per tube.
9. Spin down the cells by centrifugation at 900× *g* for 10 min at maximum acceleration and brake.

All subsequent centrifugation steps are performed at maximum acceleration and break unless specified otherwise.

10. Discard the supernatant and resuspend the pellet completely by tapping lightly against the side and bottom of the tube.
11. Add 15 mL of wash buffer to each resuspended pellet and transfer to sterile 15 mL tubes.
12. Spin down the cells by centrifugation at 400× *g* for 10 min.
13. Discard the supernatant and resuspend the pellet by tapping lightly against the side and bottom of the tube.
14. Add 10 mL of medium to one of the resuspended pellets and subsequently pool all pellets.
15. Spin down the cells by centrifugation at 350× *g* for 10 min.
16. Discard the supernatant and resuspend the pellet completely by tapping lightly against the side and bottom of the tube.
17. Resuspend the cells in 10 mL medium and count the cells. Perform extra dilution(s) as necessary.

18. Spin down the cells by centrifugation at 350× g for 10 min. Discard supernatant, resuspend the pellet completely by tapping lightly against the side and bottom of the tube and dilute the cells in medium at a concentration of 40×10^6 cells/mL.
19. Dilute the cells 1 in 1 with 2× freezing medium. For example, add 4 mL 2× freezing medium to 4 mL cell suspension.
20. Cryopreserve the cells in aliquots of 10×10^6 cells/aliquot according to standard operating procedure.

3.2 PBMC Thawing, Overnight Resting and ELISpot Plate Coating

Perform all steps in a biosafety cabinet with sterile equipment and reagents.

PBMC thawing and overnight resting

1. Pre-warm 70 mL medium per vial of cryopreserved cells to 37 °C in a 37 °C water bath.
2. Thaw a vial of cryopreserved cells in a 37 °C water bath until a small clump of ice remains.
3. Clean the outside of the cryovial with ethanol and proceed with the next steps in a biosafety cabinet.
4. Transfer cells, including ice clump, to a sterile 50 mL tube.
5. Add 10 mL of pre-warmed medium to the cells in a drip-wise fashion, allowing sufficient time between drips for the medium to disperse in the liquid already present in the tube.
6. Add another 10 mL of pre-warmed medium.
7. Spin down the cells by centrifugation at 350× *g* for 10 min at maximum acceleration and brake.

All subsequent centrifugation steps are performed at maximum acceleration and break unless specified otherwise.

8. Discard the supernatant and resuspend the pellet completely by tapping lightly against the side and bottom of the tube.
9. Add 20 mL of pre-warmed medium.
10. Spin down the cells by centrifugation at 350× *g* for 10 min.
11. Discard the supernatant and resuspend the pellet completely by tapping lightly against the side and bottom of the tube.
12. Add 1 mL of pre-warmed medium per 10×10^6 cryopreserved cells and count the cells.
13. Dilute the cells to a concentration of 2×10^6 cells/mL with pre-warmed medium.
14. Divide cells over sterile 50 mL tubes. Add a maximum of 10×10^6 cells (5 mL) per tube.
15. Put the caps on the 50 mL tubes and screw on the caps lightly. Take care to not screw the cap completely to allow for gas

exchange. Rest cells overnight at 37 °C + 5% CO_2 in a humidified incubator (*see* **Note 6**).

Coating ELISpot plate.

Perform all steps in a biosafety cabinet with sterile equipment and reagents.

1. Prepare a plate layout (*see* **Note** 7).
2. Activate the polyvinylidene difluoride (PVDF) membrane of the ELISpot plates by adding 50 μL 35% ethanol per well and incubating for 1 min at room temperature.
3. Spray off tissue paper with ethanol and place in a biosafety cabinet.
4. Remove 35% ethanol by decanting plate on tissue paper.
5. Wash the plate five times by adding 200 μL sterile water by using a multichannel pipette and decanting on tissue paper.
6. Dilute the coating antibody to a concentration of 1 μg/mL in PBS (dilute 1:1000). For coating, 100 μL is required per well: for one plate, dilute 10 μL of coating antibody into 10 mL of PBS.
7. Add 100 μL of diluted coating antibody to each well and incubate overnight at 4 °C.

3.3 T Cell Activation

Perform all steps in a biosafety cabinet with sterile equipment and reagents.

1. Spray off tissue paper with ethanol and place in a biosafety cabinet.
2. Remove the coating antibody mix from the ELISpot plate by decanting on tissue paper.
3. Wash the plate five times by adding 200 μL sterile water by using a multichannel pipette and decanting on tissue paper.
4. Add 200 μL medium/well and block the wells by incubating at room temperature for at least 30 min.
5. Obtain cells from the incubator after overnight resting. In case more than one 50 mL tube per donor was rested overnight, pool the cell suspensions per donor. Wash the 50 mL tubes by adding 10 mL medium to the first empty 50 mL tube, then transfer to the next tube to wash that tube. Once all tubes are washed, add to the tube containing the cell solution. Do this for each individual donor separately.
6. Spin down cell suspension by centrifugation at 450× *g* for 8 min.
7. Discard the supernatant and resuspend the pellet completely by tapping lightly against the side and bottom of the tube.

8. Fill up the volume to 2 mL with medium.
9. Count the cells and resuspend to a 2× concentration of the amount of cells to be tested per well (*see* **Note 8**).
10. In case cell titrations are to be tested, prepare titrations, e.g., by serial dilution.
11. Prepare cell stimulations at a 2× concentration (*see* **Notes 4** and **5**). Prepare 50 μL of stimulus per well.
12. Remove medium from the blocked and coated ELISpot plates.
13. Add 50 μL of 2× concentrated cell suspension to the coated wells.
14. Add 50 μL of 2× concentrated stimulus per well or 50 μL medium for the unstimulated wells.
15. Add 100 μL of medium to wells that are controlled for medium background.
16. Incubate plates for 24 or 48 h at 37 °C + 5% CO_2 in a humidified incubator (*see* **Notes 9** and **10**).

3.4 IFN-γ ELISpot

Step 1 should be performed in a biosafety cabinet with sterile equipment and reagents. After removing the cells from the ELISpot plate in step 1, the rest of the protocol can be performed on a lab bench outside a biosafety cabinet with clean equipment and reagents.

1. Decant the ELISpot plate to remove cells. Treat removed cells as biological waste.
2. Wash the plate(s) five times by adding 200 μL PBS by using a multichannel pipette and decanting on tissue paper. After the last wash, leave the PBS in the plate(s).
3. Prepare biotinylated detection antibody at a concentration of 1 μg/mL (dilution 1:1000) in PBS + 0.5% FCS. 100 μL is required per well: for one plate, dilute stock by adding 10 μL of biotinylated detection antibody to 10 mL of PBS + 0.5% FCS.
4. Filter detection antibody solution through a 0.2 μm filter.
5. Remove the remaining PBS from the ELISpot plate(s) by decanting.
6. Add 100 μL detection antibody solution per well and incubate at room temperature for 2 h.
7. After incubation, remove detection antibody solution by decanting.
8. Wash the plate(s) four times by adding 200 μL PBS by multichannel and decanting on tissue paper.
9. Add 200 μL PBS per well.
10. Dilute streptavidin-ALP 1 in 1000 in PBS + 0.5% FCS. 100 μL is required per well: for one plate, dilute stock 1: 1000 by adding 10 μL of streptavidin-ALP to 10 mL PBS + 0.5% FCS.

11. Remove PBS from the plate(s) by decanting.
12. Add 100 μL streptavidin-ALP solution per well and incubate at room temperature for 1 h.
13. Remove streptavidin-ALP solution from the plate(s) by decanting.
14. Wash the plate(s) four times by adding 200 μL PBS by using a multichannel pipette and decanting on tissue paper.
15. Remove the back cover from the plate(s) and wash the back of the plate(s) and the back cover(s) with demineralized water thoroughly. Once washed, dip the plate bottom once in PBS and replace the back cover.
16. Add 200 μL PBS per well.
17. Filter 5-bromo-4-chloro-3-indolyl phosphate (BCIP)/nitro blue tetrazolium (NBT) substrate solution through a 0.45 μm filter. 100 μL is required per well: for one plate, filter 10 mL of BCIP/NBIT substrate.
18. Remove PBS from the plate(s) by decanting.
19. Add 100 μL substrate solution to each well and incubate at room temperature in the dark to develop spots (*see* **Note 11**).
20. Decant the substrate solution from the plate(s) by decanting and wash the plate(s), plate bottom(s) and back cover (s) thoroughly by using tap water.
21. Allow the plate(s) to dry at room temperature for at least half an hour in front of a fan (*see* **Note 12**).
22. Store the plate(s) in the dark at room temperature until measurement (*see* **Notes 13** and **14**).

4 Notes

1. Both PBMCs isolated from whole blood and buffy coats are suitable for use in ELISpot experiments.
2. Different media can be used for ELISpot assays. Depending on the serum source, the addition of serum can influence the results (see Fig. 2). Therefore, in case rigorous serum testing is not possible, or to obtain more stable results, AIMV medium is a suitable alternative. However, authors have successfully used both AIMV medium (Figs. 1 and 4) and IMDM +5% FCS (Fig. 3).
3. Typically, cells are rested overnight to maximize viability upon the start of the assay. Authors have compared resting cells for 2 h or overnight and for the tested peptides, resting cells overnight yields optimal IFN-γ ELISpot results (Fig. 3a). Other groups have reported similar results [29].

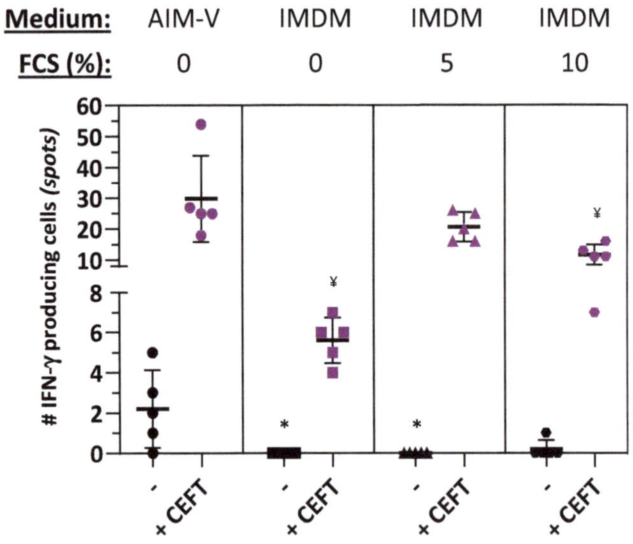

Fig. 2 Impact of medium and the addition of serum on IFN-γ ELISpot results. Cryopreserved PBMCs were thawed and rested overnight. Subsequently, 200,000 PBMCs were plated in triplicate per well and were stimulated overnight with 1 mM CEFT in indicated media overnight, or left untreated, and IFN-g producing cells were assessed by ELISpot. * indicates significantly different compared to non-stimulated AIMV ($P < 0.05$) and ¥ indicates significantly different compared to CEFT-stimulated AIMV ($P < 0.05$). Data displayed in the summary graphs are five technical replicates from one representative donor

4. As controls, authors typically stimulate cells with 10 μg/mL PHA (Merck, category # L1668) or 1 μg/mL anti-CD3 + 1 μg/mL anti-CD28 (e.g., included in ELISpot kit) (see Figs. 1c and 4a). For antigen-specific stimulation, authors have successfully employed 1 μM CEFT (JPT #PM-CEFT), 1 μM SARS-Cov-2 nucleocapsid peptide pool (JPT, category #PM-WCPV-NCAP), 1 μM SARS-CoV-2 Spike peptide pool 1 (JPT #PM-WCPV-S-1, pool 1), or 1 μM SARS-CoV-2 peptide pool 2 (JPT #PM-WCPV-S-1, pool 2) for human cells (see Fig. 1b, d), or 0.5 mg/mL $OVA_{323-339}$ for murine cells [20]. For vaccine candidate peptides, typically, between 0.625 μg/mL and 10 μg/mL is employed [24, 30–32].
5. Memory T cells, in particular $CD8^+$ T cells, generally do not require co-stimulatory signals for the production of cytokines [7, 12]. Indeed, when we compared whether the addition of 1 μg/mL anti-CD28 (Sanquin, clone 15E8) and/or anti-CD29 (BioLegend, clone TS2/16) in addition to peptide stimuli increased the window of detection in our T cell ELISpot assay, when taking into account the higher background signal induced by the co-stimulatory antibodies, this resulted in only

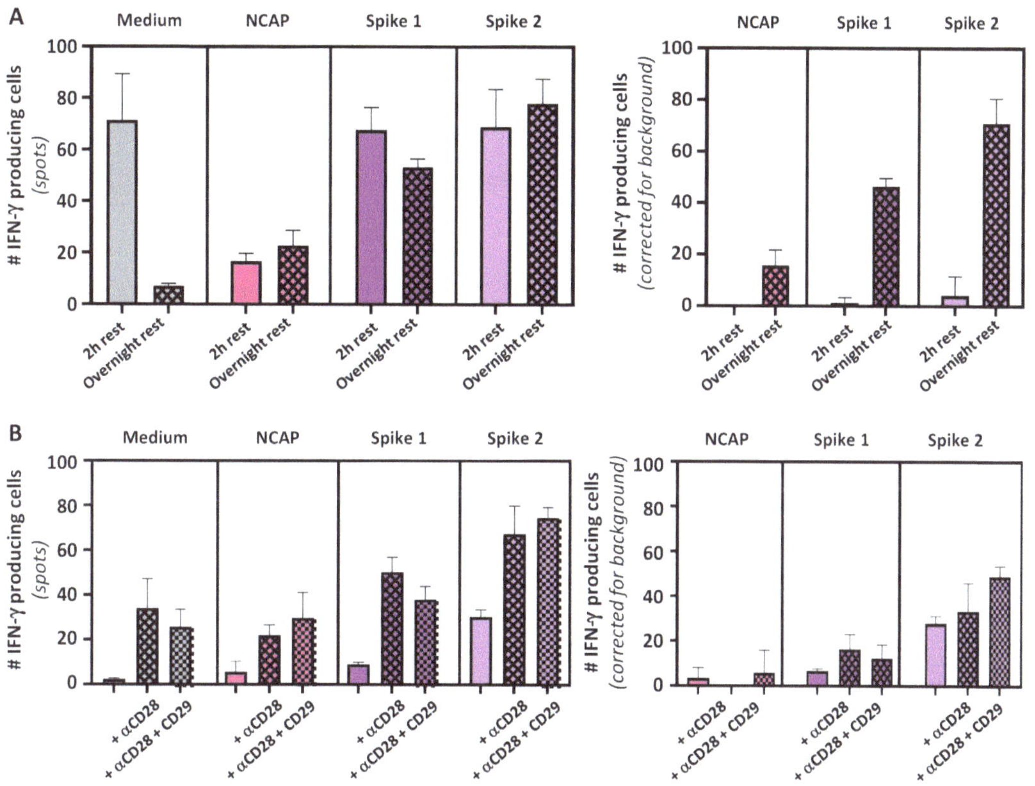

Fig. 3 Resting cells overnight and not adding co-stimulatory molecules reduces background IFN-γ production by human PBMCs. (**a**) Cryopreserved PBMCs were thawed and allowed to rest for 2 h or overnight. Subsequently, 200,000 PBMCs were plated per well and stimulated overnight with 1 mM of indicated SARS-CoV-2 peptides in IMDM +5% FCS. Left, summary graph of number of IFN-γ producing cells [spots], right, summary graph of IFN-γ producing cells [corrected for background IFN-γ production]. Both graphs display the mean ± SD of three technical replicates of one representative healthy donor out of two donors. (**b**) Cryopreserved PBMCs were thawed and allowed to rest overnight. Subsequently, 200,000 PBMCs were plated per well and left untreated, or stimulated overnight with 1 mM of indicated SARS-CoV-2 peptides, with or without the addition of 1 mg/mL anti-CD28 or 1 mg/mL anti-CD28 + anti-CD29, all in IMDM +5% FCS. Left, summary graph of number of IFN-γ producing cells [spots] and right, summary graph of IFN-γ producing cells [corrected for background IFN-γ production]. Both graphs display the mean ± SD of three technical replicates of one representative healthy donor out of two donors

marginal increases, if any (Fig. 3b). Therefore, in order to minimize assay background, when using high affinity peptides such as those employed in this assay (SARS-CoV-2 nucleocapsid, spike pool 1, and spike pool 2, respectively), the addition of co-stimulatory molecules is not required. However, this should be experimentally determined for every individual peptide.

6. When using cryopreserved cells, depending on the laboratory, some researchers rest cells, either shortly or overnight, or immediately proceed to cellular stimulation. In our experience, not resting cells overnight results in increased background

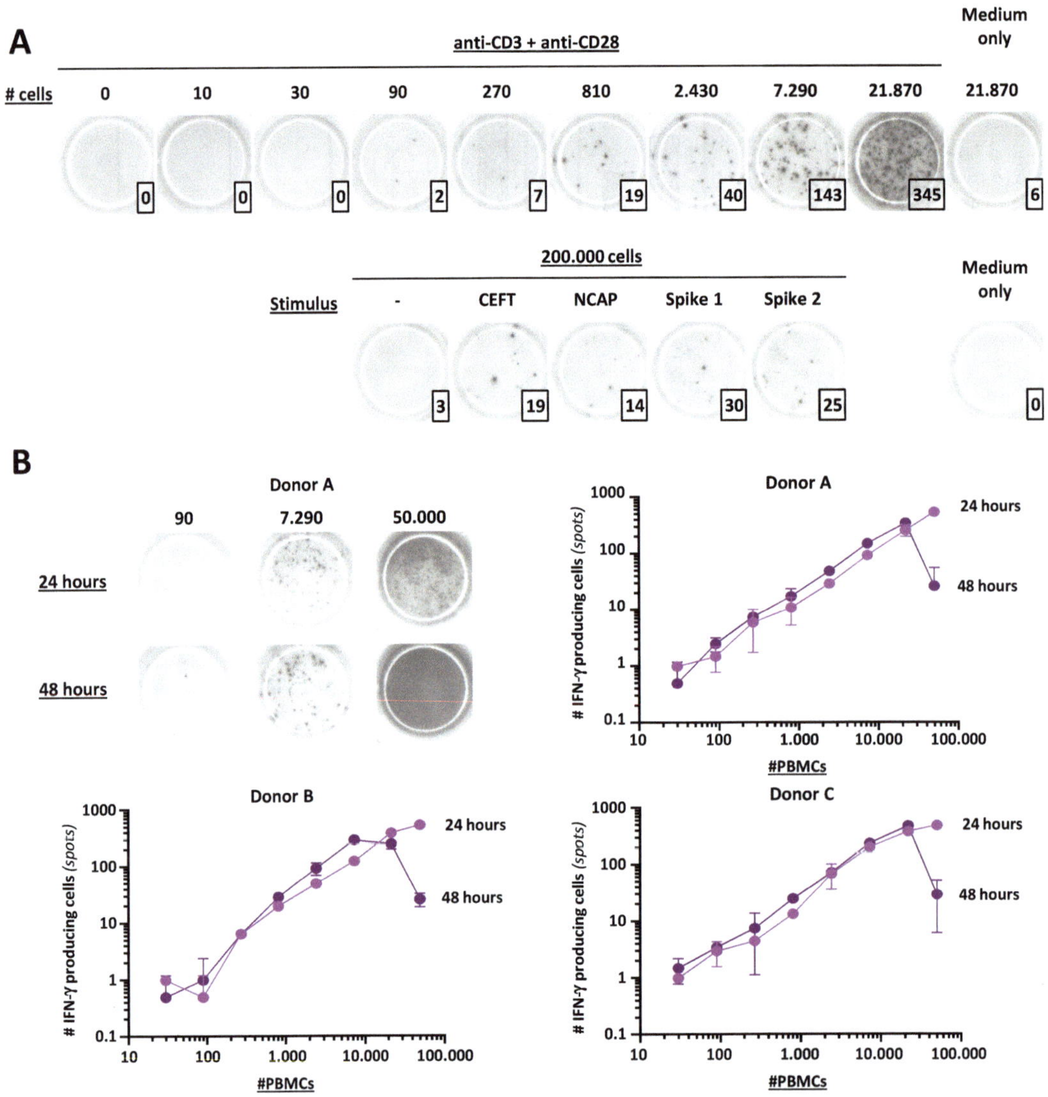

Fig. 4 Impact of cell numbers and time of incubation on IFN-γ ELISpot limit of detection. Indicated amounts of cryopreserved PBMCs were thawed, allowed to rest overnight, and plated in triplicate per well. Cells were subsequently stimulated for indicated time with indicated stimuli, or left untreated, all in AIMV medium. IFN-γ producing cells were assessed by ELISpot. (**a**). Cells were stimulated overnight with 1 mg/mL anti-CD3 + 1 mg/mL anti-CD28 (top row) or 1 mM of indicated stimuli (bottom row). Numbers above the image indicate the number of plated cells, and numbers in the image indicate counts of IFN-γ producing cells. (**b**) Indicated amounts of cryopreserved PBMCs were stimulated for 24 (light colors) or 48 (dark colors) hours with 10 mg/mL PHA. Data displayed in summary graphs are mean ± SD of three technical replicates of three healthy donors, representing at least two independently performed experiments

stimulation (Fig. 3a). A high background production of IFN-γ makes data interpretation difficult, as correcting for the background will result in no IFN-γ production upon stimulation when resting only shortly. Therefore, authors suggest resting cells overnight prior to inclusion into ELISpot experiments.

7. When preparing a plate layout, the following parameters should at least be considered: amount of replicates, experimental controls, and technical controls. In general, authors employ at least two, but preferably three replicates per condition to be tested. However, in case low frequencies of responding cells are expected, the number of replicates should be increased. As experimental controls, authors often employ PHA or anti-CD3 + anti-CD28 (*see* **Note 4,** Figs. 1c and 4a). Of note, PHA also serves as a technical control. Other technical controls the authors employ are wells with only medium and unstimulated cells plated at the highest density employed for experimental conditions. In the literature, up to six replicates have been previously advised [22].
8. The appropriate number of cells per stimulation should be experimentally determined. In general, for controls such as PHA and anti-CD3 + anti-CD28, authors plate 10,000 cells (see Figs. 1c, 3b and 4a). The amount of responding cells should be in the linear detection range. Especially for pan-T cell stimuli, the linear range is narrow. For example, for PHA stimulation is linear up to at least 21,870 cells for 24-h stimulations, and between 7920 and 21,870 cells for 48-h stimulations (Table 1). For antigen-specific responses, in general, 200,000 cells are sufficient (see Fig. 4a). However, this should be experimentally confirmed. Please note that a maximum of 300,000 PBMCs per well is advised to ensure the formation of a monolayer. When more cells are added per well, PBMCs will not be able to form a monolayer, which can negatively impact on results.
9. Once the cells have been plated, it is of extreme importance to not disturb the plate during incubation. This is in order to prevent already settled cells that have secreted cytokines from moving, resulting in spot duplication or measurement artifacts. Therefore, the location of incubation should be carefully considered. In general, authors prefer to incubate in the back of an incubator that is not used by other experimenters, away from any mechanical devices (e.g., centrifuges).
10. Depending on the stimuli, the duration of stimulation might vary. However, for anti-CD3 + anti-CD28 (data not shown) or for PHA stimulation (Fig. 4b), both 24 and 48 h of stimulation provide sufficient activated T cells. However, as indicated in Table 1, for polyclonal stimulations, the linear range of the IFN-γ ELISpot is between 0 and 21,870–50,000 cells for 24-h stimulation for all three donors, but for 48-h stimulations, it is linear only when using up to 7920 cells per well. This is also evident from the high signal observed upon visual inspection of wells with 50,000 cells incubated for 48 h (Fig. 4b). Linearity should be experimentally validated for polyclonal stimuli. In

Table 1
Linearity of IFN-γ ELISpot upon PHA stimulation.

Stimulation time	24 h				48 h			
# of cells	7920	**21,870**	50,000	100,000	**7920**	21,870	50,000	100,000
R^2 donor A	0.9805	**0.9647**	0.9908	0.891	**0.9953**	0.9771	0.06859	0.000394
R^2 donor B	0.9773	**0.9971**	0.9342	0.3114	**0.9886**	0.6032	0.01691	0.006982
R^2 donor C	0.9757	**0.9568**	0.8677	0.2181	**0.9963**	0.971	0.05717	0.002899

general, for low-frequency antigenic responses, linearity should not be an issue.

11. Monitor spot development carefully. Usually, spots will develop within 10 min after starting incubation.
12. This step can also be performed without a fan but will result in longer drying times.
13. In the authors' experience, spots continue to develop, and optimal results are obtained when waiting at least 20 h after finalizing staining before measuring ELISpot plates.
14. For spot measurement, authors employ an AID iSpot Spectrum ELISpot reader and use standard settings for IFN-γ. Depending on the antigen used, exposure time, spot size, and spot intensity settings should be optimized.

References

1. Almeida JR, Price DA, Papagno L et al (2007) Superior control of HIV-1 replication by CD8+ T cells is reflected by their avidity, polyfunctionality, and clonal turnover. J Exp Med 204:2473–2485. https://doi.org/10.1084/jem.20070784
2. Ciuffreda D, Comte D, Cavassini M et al (2008) Polyfunctional HCV-specific T-cell responses are associated with effective control of HCV replication. Eur J Immunol 38:2665–2677. https://doi.org/10.1002/eji.200838336
3. Wang P, Gu Y, Zhang Q et al (2012) Identification of resting and type I IFN-activated human NK cell miRNomes reveals MicroRNA-378 and microRNA-30e as negative regulators of NK cell cytotoxicity. J Immunol 189:211–221. https://doi.org/10.4049/jimmunol.1200609
4. Pennock ND, White JT, Cross EW et al (2013) T cell responses: naïve to memory and everything in between. Adv Phsiol Educ 37:273–283. https://doi.org/10.1152/advan.00066.2013
5. Voskoboinik I, Whisstock JC, Trapani JA (2015) Perforin and granzymes: function, dysfunction and human pathology. Nat Rev Immunol 15:388–400. https://doi.org/10.1038/nri3839
6. Golubovskaya V, Wu L (2016) Different subsets of T cells, memory, effector functions, and CAR-T immunotherapy. Cancers (Basel) 8. https://doi.org/10.3390/cancers8030036
7. Salerno F, Freen-van Heeren JJ, Guislain A et al (2019) Costimulation through TLR2 drives polyfunctional $CD8^+$ T cell responses. J Immunol 202:714–723. https://doi.org/10.4049/jimmunol.1801026
8. Freen-van Heeren JJ, Popović B, Guislain A, Wolkers MC (2020) Human T cells employ conserved AU-rich elements to fine-tune IFN-γ production. Eur J Immunol 50:949–958. https://doi.org/10.1002/eji.201948458
9. Oja AE, Saris A, Ghandour CA et al (2020) Divergent SARS-CoV-2-specific T- and B-cell responses in severe but not mild COVID-19 patients. Eur J Immunol 50:1998–2012. https://doi.org/10.1002/eji.202048908
10. Finco D, Grimaldi C, Fort M et al (2014) Cytokine cytokine release assays: current practices and future directions. Cytokine 66:143–155. https://doi.org/10.1016/j.cyto.2013.12.009
11. Freen-van Heeren JJ, Nicolet BP, Wolkers MC (2018) Measuring T cell responses by flow cytometry–based fluorescence in situ hybridization. Crit Rev Immunol 38:131–143. https://doi.org/10.1615/CritRevImmunol.2018025938
12. Oja AE, Piet B, van der Zwan D et al (2018) Functional heterogeneity of CD4+ tumor-infiltrating lymphocytes with a resident memory phenotype in NSCLC. Front Immunol 9:2654. https://doi.org/10.3389/FIMMU.2018.02654
13. Cossarizza A, Chang HD, Radbruch A et al (2019) Guidelines for the use of flow cytometry and cell sorting in immunological studies (second edition). Eur J Immunol 49:1457–1973. https://doi.org/10.1002/eji.201970107

14. Freen-van Heeren JJ, Nicolet BP, Wolkers MC (2020) Combined single-cell measurement of cytokine mRNA and protein in immune cells. Methods Mol Biol 2108:259–271
15. Körber N, Behrends U, Hapfelmeier A et al (2016) Validation of an IFNγ / IL2 FluoroSpot assay for clinical trial monitoring. J Transl Med 1–12. https://doi.org/10.1186/s12967-016-0932-7
16. Shafer-Weaver K, Sayers T, Strobl S et al (2003) The Granzyme B ELISPOT assay: an alternative to the 51Cr-release assay for monitoring cell-mediated cytotoxicity. J Transl Med 1:1–9. https://doi.org/10.1186/1479-5876-1-14
17. Sundararaman S, Karulin A, Ansari T et al (2015) High reproducibility of ELISPOT counts from nine different laboratories. Cell 4:21–39. https://doi.org/10.3390/cells4010021
18. Slota M, Lim JB, Dang Y, Disis ML (2011) ELISpot for measuring human immune responses to vaccines. Expert Rev Vaccines 10: 299–306. https://doi.org/10.1586/erv.10.169
19. Malyguine AM, Strobl S, Dunham K et al (2012) ELISPOT assay for monitoring cytotoxic T lymphocytes (CTL) activity in cancer vaccine clinical trials. Cell 1:111–126. https://doi.org/10.3390/cells1020111
20. ten Brinke A, Marek-Trzonkowska N, Mansilla M et al (2017) Monitoring T-cell responses in translational studies: optimization of dye-based proliferation assay for evaluation of antigen-specific responses. Front Immunol 8:1870. https://doi.org/10.3389/fimmu.2017.01870
21. Helms T, Boehm BO, Asaad RJ et al (2000) Direct visualization of cytokine-producing recall antigen-specific CD4 memory T cells in healthy individuals and HIV patients. J Immunol 164:3723–3732. https://doi.org/10.4049/jimmunol.164.7.3723
22. Janetzki S, Price L, Schroeder H et al (2015) Guidelines for the automated evaluation of Elispot assays. Nat Protoc 10:1098–1115. https://doi.org/10.1038/nprot.2015.068
23. Sahin U, Muik A, Derhovanessian E et al (2020) COVID-19 vaccine BNT162b1 elicits human antibody and TH1 T cell responses. Nature 586. https://doi.org/10.1038/s41586-020-2814-7
24. Woldemeskel BA, Garliss CC, Blankson JN et al (2021) SARS-CoV-2 mRNA vaccines induce broad CD4+ T cell responses that recognize SARS-CoV-2 variants and HCoV-NL63. J Clin Invest 131
25. Schmitz J, Thiel A, Rajewsky K et al (1994) Induction of interleukin 4 (IL-4) expression in T helper (Th) cells is not dependent on IL-4 from non-Th cells. Cell 179:1349–1353
26. Filbey KJ, Camberis M, Chandler J et al (2019) Intestinal helminth infection promotes IL-5- and CD4 + T cell-dependent immunity in the lung against migrating parasites. Mucosal Immunol 12:352–362. https://doi.org/10.1038/s41385-018-0102-8
27. Dillenbeck T, Gelius E, Fohlstedt J, Ahlborg N (2014) Triple cytokine FluoroSpot analysis of human antigen-specific IFN-γ, IL-17A and IL-22 responses. Cell 3:1116–1130. https://doi.org/10.3390/cells3041116
28. Janetzki S, Rueger M, Dillenbeck T (2014) Stepping up ELISpot: multi-level analysis in FluoroSpot assays. Cell 3:1102–1115. https://doi.org/10.3390/cells3041102
29. Santos R, Buying A, Sabri N et al (2014) Improvement of IFNγ ELISPOT performance following overnight resting of frozen PBMC samples confirmed through rigorous statistical analysis. Cell 4:1–18. https://doi.org/10.3390/cells4010001
30. Dubey S, Clair J, Fu TM et al (2007) Detection of HIV vaccine-induced cell-mediated immunity in HIV-seronegative clinical trial participants using an optimized and validated enzyme-linked immunospot assay. J Acquir Immune Defic Syndr 45:20–27. https://doi.org/10.1097/QAI.0b013e3180377b5b
31. Frahm N, DeCamp AC, Friedrich DP et al (2012) Human adenovirus-specific T cells modulate HIV-specific T cell responses to an Ad5-vectored HIV-1 vaccine. J Clin Invest 122:359–367. https://doi.org/10.1172/JCI60202
32. Aruga A, Takeshita N, Kotera Y et al (2013) Long-term vaccination with multiple peptides derived from cancer-testis antigens can maintain a specific T-cell response and achieve disease stability in advanced biliary tract cancer. Clin Cancer Res 19:2224–2231. https://doi.org/10.1158/1078-0432.CCR-12-3592

Correction to: Immune Homeostasis: A Novel Example of Teamwork

Vijay Kumar and John H. Stewart IV

Correction to:
Chapter 1 in: Vijay Kumar (ed.), *Immune Homeostasis: Methods and Protocols*, Methods in Molecular Biology, vol. 2782,
https://doi.org/10.1007/978-1-0716-3754-8_1

In Chapter 1 of the original version of this book, the header of Table 1 had a mistake. The term "TLRs" needs to be replaced with "PRRs". This has been rectified as follows:

Table 1
Different PRRs, their localization, and major ligands

PRRs	Localization	Ligands (PAMPs and DAMPs)	Origin of ligands
TLR1	Plasma membrane	Triacyl lipopeptide Soluble factors	Bacteria and mycobacteria
TLR2	Plasma membrane and endosomes	Peptidoglycan (PGN), lipoteichoic acid (LTA), lipoproteins or lipopeptides, lipoarabinomannan, glycolipids, porins, zymosan, atypical LPS, heat shock protein 70 (Hsp70), and eosinophil-derived neurotoxin (EDN)	Gram positive bacteria, mycobacteria, *S. epidermidis*, *Trypanosoma cruzi*, *Treponema maltophilum*, *Neisseria*, *Fungi*, *Leptospira interrogans*, *Porphyromonas gingivalis*, and host

(continued)

The updated version of this chapter can be found at
https://doi.org/10.1007/978-1-0716-3754-8_1

Vijay Kumar (ed.), *Immune Homeostasis: Methods and Protocols*, Methods in Molecular Biology, vol. 2782,
https://doi.org/10.1007/978-1-0716-3754-8_18,

Table 1 (continued)

PRRs	Localization	Ligands (PAMPs and DAMPs)	Origin of ligands
TLR3	Endoplasmic reticulum (ER), endosomes, multivesicular bodies, lysosomes, and endolysosomes	dsRNA and ssRNA and synthetic analog polyinosinic-polycytidylic acid (poly I:C)	dsRNA, ssRNA, and dsDNA viruses
TLR4	Plasma membrane and endosome	LPS, Taxol, fusion protein, envelope proteins, high mobility group box 1 protein (HMG-B1), Hsp60, Hsp70, Hsp22, Hsp96 type III repeat extra domain A of fibronectin, hyaluronic acid, heparin sulfate, fibrinogen, saturated fatty acids, and fetuin-A	Gram negative bacteria, plant, respiratory syncytial virus (RSV), mouse mammary tumor virus (MMTV), *Chlamydia pneumoniae*, *Chlamydia trachomatis,* and host
TLR5	Plasma membrane	Flagellin	Bacteria
TLR6	Plasma membrane	Di-acyl lipopeptides and zymosan	Mycoplasma
TLR7	ER, endosomes, multivesicular bodies, lysosomes, and endolysosomes	Guanosine- and uridine-rich ssRNA, loxoribine (a guanine analog), bropirimine, ribonucleoproteins (RNPs), siRNAs, and imidazoquinoline derivatives such as resiquimod (R848) and imiquimod	Viruses (human immunodeficiency virus-1 or HIV-1, influenza virus and vesicular stomatitis virus or VSV), and synthetic compounds
TLR8	ER, endosomes, multivesicular bodies, lysosomes, and endolysosomes	ssRNA and RNPs	Viruses
TLR9	ER, endosomes, multivesicular bodies, lysosomes, and endolysosomes	CpG oligodeoxyneucleotide (ODN), viral double-stranded DNA (dsDNA), and hemozoin pigment	Bacteria and viruses (Herpes Simplex virus-1 and -2 or HSV-1 and -2, mouse cytomegalovirus or MCMV), malaria, and protozoa genome (*Trypanosoma cruzi*)

(continued)

Table 1 (continued)

PRRs	Localization	Ligands (PAMPs and DAMPs)	Origin of ligands
TLR10 (humans)	ER, endosomes, multivesicular bodies, lysosomes, and endolysosomes	dsRNA [90]	Viruses
TLR11	ER, endosomes, multivesicular bodies, lysosomes, and endolysosomes	Profilin-like protein	*Toxoplasma gondii*
TLR12	ER, endosomes, multivesicular bodies, lysosomes, and endolysosomes	Profilin-like protein	*Toxoplasma gondii*
TLR13 (mouse)	ER, endosomes, multivesicular bodies, lysosomes, and endolysosomes	23s ribosomal RNA	Bacteria
NOD-like receptor (NLR) family			
NLRA family member			
MHC class II transactivator or CIITA	Nucleus of macrophages, dendritic cells (DCs), B cells, and T cells, drives MHC II transcription and other genes critical for antigen presentation, including invariant chain and HLA-DM genes	IFN-γ is a CIITA activator	Viral infections
NLRB family member			
Neuronal apoptosis inhibitory protein or NAIP (in humans) and NAIP1-7 (in mice)	Cytosol of myeloid immune cells (MICs) and forms inflammasome-associated with NLRC4	Flagellin, needle component of type 3 secretory system (T3SS) of the bacteria	*Legionella pneumophila, Salmonella typhimurium, Shigella flexneri,* and *Pseudomonas* spp.
NLRC family members			
Nucleotide-binding oligomerization domain 1 (NOD1)	Cytosol and activates NF-κB	Bacterial peptides, including L-Ala-γ-D-Glu-*meso*-diaminopimelic acid (Tri-DAP)	Gram-negative and -positive bacteria

(continued)

Table 1 (continued)

PRRs	Localization	Ligands (PAMPs and DAMPs)	Origin of ligands
NOD2	Cytosol, activates NF-κB, autophagy, and type 1 IFN production	Muramyl dipeptide (MDP)	Gram-negative and -positive bacteria
NLR family CARD domain–containing 3 (NLRC3 or sNOD3)	Cytosol of immune cells, negative regulator of T cell activation, and lipopolysaccharide (LPS)-induced immune activation, including NF-κB activation, stimulator of interferon genes (STING), and TANK-binding kinase 1 (TBK1) signaling pathway	Recognizes viral DNA and other nucleic acids via its leucine-rich repeat (LRR) domain	Viruses
NLRC4	Cytosol, plays a role in caspase 1 activation, cell death, and phagosome maturation, and works downstream to NAIP	Rod proteins and flagellin	Bacteria
NLRC5	Shuttles between nucleus and cytosol, expressed constitutively in NK, NKT, T, and B cells, and primarily upregulates MHC I and accessory protein gene expression	IFN-γ and IFN-β are NLRC5 activators. Also, NLRP3 inflammasome activators (pathogens) overlap with NLRC5 inflammasome activation	Bacteria
NLR family member X 1 or NLRX1	Ubiquitous in mammals and located in the outer mitochondrial membrane and mitochondrial matrix, critical for reactive oxygen species (ROS) production, regulates antiviral immunity, autophagy, and mitophagy	Recognizes Beclin 1 and human papillomavirus 16 (HPV16) E7 protein	Viruses

(continued)

Table 1 (continued)

PRRs	Localization	Ligands (PAMPs and DAMPs)	Origin of ligands
NLRP family members			
NLR family pyrin-domain containing 1 or NLRP1 or CARD1 or NALP1 (first PRR discovered to form inflammasome	In humans, keratinocytes and epithelial cells, including lung epithelial cells, highly express NLRP1 in the nucleoplasm and cytosol. Motor neurons of cerebral cortex and spinal cord, seminiferous tubules, and microglia, along with other immune cells	Human NLRP1 recognizes muramyl dipeptide (MDP), viral proteases, including SARS-CoV-2 and dsRNA, whereas murine NLPRPB1 recognizes virulence factor lethal toxin (LT) of the *Bacillus anthracis*	*Toxoplasma gondii* (a parasite), viruses, and bacteria
NLRP2 or NALP2	Localized to the Golgi apparatus (GA), vesicles, and cytosol of immune cells, including astrocytes, suppresses NF-κB activity, and suppresses HLA-C expression in human trophoblasts	IFN, adenosine triphosphate (ATP), and LPS can activate NLRP2	Bacteria
NLRP3 or NALP3	Inactivated NLRP3 resides in the endoplasmic reticulum (ER) and cytosol, upon activation NLRP3 inflammasome assembles in the cytosol, mitochondria, and mitochondria-associated ER membranes. It may also localize on or adjacent to GA to form an inflammasome. Myeloid immune cells (MICs) highly express NLRP3. T cells, B cells, endothelial cells, epithelial cells, and fibroblasts also express NLRP3	Several self-DAMPs, MAMPs, PAMPs, and external molecules, such as alum, silica, aluminum hydroxide, nanoparticles, UVB rays, imiquimod, and resiquimod	Host, bacteria, viruses, and fungi

(continued)

Table 1 (continued)

PRRs	Localization	Ligands (PAMPs and DAMPs)	Origin of ligands
NLRP4 or NALP4	Localized to mitochondria, and not detected in immune cells but highly expressed in the placenta, testes, spleen, pancreas, liver, lungs, kidneys, and thymus. It negatively regulates TNF-α and IL-1β-induced NF-κB activation, dsDNA and dsRNA-induced type 1 IFN signaling, and regulates autophagy during bacterial infections	ATP and nucleotides	Host and viruses
NLRP5	Localized in the GA and cytosolic vesicles, not detected in immune cells and human brain but expressed in ovaries, oocytes, testes (spermatogonia), and parathyroid gland, critical for embryogenesis	ATP and nucleotides	Host
NLRP6 or NLRC6 or PYPAF5 or PAN3	Cytosol of granulocytes, T cells, monocytes, macrophages, DCs, enterocytes or intestinal epithelial cells (IECs), and myofibroblasts. NLRP6 activation inhibits NF-κB activation and caspase 1 activation	Senses microbiota metabolites, such as taurine, LTA, and LPS. However, it directly interacts with the viral dsRNA and forms a dynamic liquid-like phase condensate or liquid-liquid phase separation (LLPS)	Bacteria and viruses
NLRP7 or NALP7 or PYPAF3 or PAN7 or NOD12 (only found in primates)	Localized to GA of the cells of reproductive tract (testes – spermatogonia, spermatocytes, and late spermatids) and ovaries – oocytes),	ATP, LPS, Pam3CSK4, PGN, LTA, and microbial acylated lipopeptides	Gram-positive bacteria, *Mycoplasma* spp. and *Mycobacterium bovis*

(continued)

Table 1 (continued)

PRRs	Localization	Ligands (PAMPs and DAMPs)	Origin of ligands
	extravillous trophoblasts, plasma cells, plasmacytoid DCs, macrophages, and liver NK cells. Inhibits caspase-1-depedent IL-1β secretion		
NLRP9 or NALP9 or NOD6, and PAN12 (Human) and NLRP9a, b, and c isoforms are present in mice	Expressed at the inner cell membrane of intestinal cells and reproductive organs such as ovaries, oocytes, and testes. Not found in immune cells.	Short dsRNA	Viruses
NLRP10 or NALP10 or PAN5 or NOD8 or PYNOD	Localized to nucleoplasm and cytosol of cells of a variety of tissue, with a high expression in the heart, skeletal muscles, brain, and skin (keratinocytes), but not in immune cells. It is also expressed in distal colonic epithelial cells (CECs). It inhibits caspase-1-dependent IL-1β secretion. It also exhibits ATPase and GTPase activity. It lacks ligand binding LRR domain. Critical for regulating host adaptive immunity during *Candida albicans* infection and other inflammatory conditions such as gut autoinflammation and skin inflammatory diseases	ATP, damaged mitochondria (in response phospholipase C activator-3m3-FBS) independent of mitochondrial DNA (mtDNA)	Host
NLRP12 or NALP12 or PAN6 or PYPAF7 or Monarch1	Localized to the cytosol of most MICs and highly expressed in gastrointestinal cells. It	ATP, heme + PAMPs or TNF, and *Yersinia pestis*	Host and pathogens

(continued)

Table 1 (continued)

PRRs	Localization	Ligands (PAMPs and DAMPs)	Origin of ligands
	inhibits NF-κB and ERK activation pathways. Also, acts as a negative regulator of NOD2 and RIG-1 (RLR) signaling pathway		
NLRP14 or NALP14 or NOD5 or PAN8	Expressed in cytosol of testes, ovarian follicle cells, and oocytes, but not in immune cells. Important for spermatogenesis and fertilization by negatively regulating cGAS/STING and RLR-mediated downstream signaling pathways through binding TBK1	Recognizes TBK1	Host protein
Absent in melanoma-2-like receptor (ALR) AIM-2	Cytosol of hematopoietic cells and its activation produces caspase-1-dependent IL-1β and IL-18 generation and pyroptosis	Recognizes cytosolic dsDNA	Host and pathogens
cGAS/STING signaling pathway	Cytosol and nucleus, but only cytosolic cGAS acts as a PRR, and STING is an adaptor molecule that recognizes cGAMP or CDNs to generate TBK1-dependent NF-κB and IRF-3-mediated type 1 IFN and cytokine release	cGAS recognizes cytosolic dsDNA	Host and pathogens
RIG-1-like receptors (RLRs)	Cytosol of immune cells and most other cells. RLR activation synthesizes IRF3/IRF7/NF-κB-dependent cytokines and IFNs	Recognizes viral DNA and RNA	Viruses

INDEX

Vijay Kumar (ed.), *Immune Homeostasis: Methods and Protocols*, Methods in Molecular Biology, vol. 2782, https://doi.org/10.1007/978-1-0716-3754-8,